# Annotated Checklist of the VASCULAR PLANTS of Grand Teton National Park and Teton County, Wyoming

by Richard J. Shaw, Ph. D.
Professor Emeritus
and Emeritus Director of Intermountain Herbarium,
Department of Biology,
Utah State University, Logan, Utah

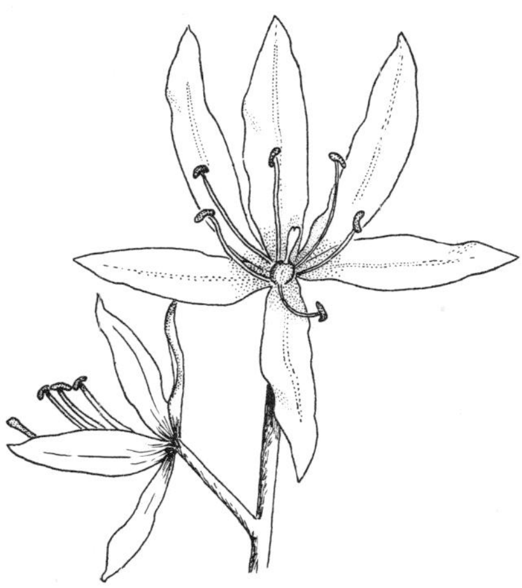

Grand Teton Natural History Association
Grand Teton National Park
Moose, Wyoming
1992

## Acknowledgments

Funding for the publication of this checklist was provided by the Grand Teton Natural History Association. I wish to express my sincere appreciation to all who helped in the project and especially Sharlene Milligan of the association who gave encouragement and careful editing. Jill Hodges and Jackie Gilmore performed the tedious task of word processing and Hollis Marriott of the WYNDD, Nature Conservancy very generously provided information from her field surveys of rare plants. I am also grateful to Dr. Mary Barkworth, Director of the Intermountain Herbarium, for her general consultation and specific advice on the Poaceae.

Richard J. Shaw

## Illustrations Credits

Front and back cover illustrations by Tina Close
Front cover: *Campanula rotundifolia,* Scotch Bluebell, Harebell
Back cover: *Fragaria virginiana,* Virginia Strawberry

Black and white illustrations by Richard J. Shaw
Title page: *Camassia quamash,* Blue Camas
Page iii: *Parnassia fimbriata,* Rocky Mt. Parnassus
Page 74: *Epilobium angustifolium,* Fireweed
Page 76: *Hyoscyamus niger,* Black Henbane
Page 77: *Dicentra uniflora,* Steershead



Design and production by Jackie Gilmore, Backwaters Publications, Moose, Wyoming
Printed by Lorraine Press, Salt Lake City, Utah

ISBN 0–931895–23–5

# Table of Contents



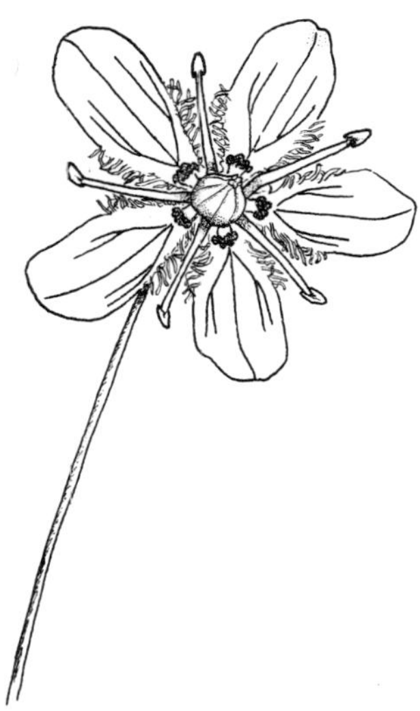

## Introduction

This checklist was assembled for several reasons: (1) my *Field Guide to the Vascular Plants of Grand Teton National Park and Teton County, Wyoming* which appeared in 1976 has long been out of print; (2) over 125 additional species have been found in the park and county that were not recognized in 1976; (3) tremendous changes have occurred in roads and trails within the park in the last sixteen years as well as a startling increase in population in the county; (4) a list is needed which summarizes the known information of the flora for visitors, students and vegetation managers.

The following list was compiled using literature published since 1952 and pressed specimens at the following herbaria: GTNP Herbarium, Moose, Wyoming, Intermountain Herbarium at Utah State University, Logan, Utah, and Rocky Mountain Herbarium at the University of Wyoming, Laramie, Wyoming. These data were supplemented with the author's personal experiences of collecting in Grand Teton National Park for the past forty-one years. Each entry is followed by the most frequently used common name, habitat information, flowering period and special comments such as place of origin for alien species, species of concern, and poisonous species. Only the most recent synonym(s) are given for taxa found in selected literature. Family nomenclature follows Cronquist (1981); generic and specific epithet nomenclature for the most part follows Dorn (1988). Species are listed in phylogenetic order by division: Lycopodiophyta, Equisetophyta, Polypodiophyta, Pinophyta and Magnoliophyta and by class Magnoliopsida and Liliopsida within Magnoliophyta. Within these groups species are listed alphabetically by family, genus and species. There is no doubt that the checklist is incomplete—an inherent problem in any catalog of plant species. Nevertheless, it is important that an attempt be made to summarize plant data as we now understand it in the last decade of the twentieth century. The author will appreciate receiving any changes in data about the plant taxa included in this checklist.

† precedes known aliens in the species listing.

* precedes all species of concern in the species listing.

## Aliens in the Flora

The word "alien" in this catalog is used to refer to those plants that have migrated directly into the park and adjacent county areas within the last hundred years, or those plants remaining from cultivation, usually around abandoned places of habitation. The status of such aliens is difficult to determine. Some floristic botanists feel that a checklist should include only well established aliens, but exclude those recorded only once or twice. The author feels, however, that in view of the increasing visitation to Grand Teton National Park and the entire Greater Yellowstone Ecosystem that an attempt should be made to document as many aliens as possible. Changes in the status of aliens have occurred, and their role in the flora should be periodically reevaluated. Comparison of the 1968 checklist with this list will point to an increasing "pollution" of the park and the county with aliens some of which have already become noxious weeds. In 1968, 72 aliens were noted in Shaw's checklist. In 1976, Shaw's Field Guide listed 88 aliens. In this checklist aliens have increased to 117, a percentage increase of 63% since 1968. This clearly should be disturbing to vegetation managers. The symbol "†" precedes a known alien in the species listing.

## DIVISION LYCOPODIOPHYTA CLUBMOSSES

### Isoetaceae Quillwort Family

*Isoetes bolanderi* Engelm.
Bolander Quillwort
Rare in shallow water of ponds near Colter Bay Village.

### Lycopodiaceae Clubmoss Family

*Lycopodium annotinum* L.
Stiff Clubmoss
Deep shade on the west sides of Jenny and Leigh Lakes; circumboreal.

* *Lycopodium selago* L.
Fir Clubmoss
Rare, occurring in cliffs. Only one collection. (Rocky Mt. Herbarium)

### Selaginellaceae Spikemoss Family

*Selaginella densa* Rydb.
Rydberg Spikemoss
Exposed rocky sites in the valley and sometimes extending above tree line.

* *Selaginella selaginoides* (L.) Link
Low Selaginella
Rare, Treasure Lake, Teton Canyon.

## DIVISION EQUISETOPHYTA HORSETAILS

### Equisetaceae Horsetail Family

*Equisetum arvense* L.
Meadow Horsetail
Moist areas of the valley; circumboreal.

* *Equisetum fluviatile* L.
Water Horsetail
Marshes and bogs. Only one collection made about 1900. (Rocky Mt. Herbarium)

*Equisetum hyemale* L. var. *affine* (Engelm.) Eaton
Common Scouringrush
Area of Oxbow Bend of the Snake River; circumboreal.

*Equisetum laevigatum* A. Br.
Smooth Scouringrush
Sedge meadows near the Oxbow Bend of the Snake River.

*Equisetum variegatum* Schleich. ex Weber & Mohr
Variegated Scouringrush
Wet meadows and stream banks of the valley; circumboreal.

## DIVISION POLYPODIOPHYTA FERNS

### Marsileaceae Pepperwort Family

*Marsilea vestita* Hook. & Grev. var. *oligospora* (Goodding) Dorn
Pepperwort
Small ponds of the valley. Sporocarps require at least 2 years to mature.

## Ophioglossaceae Adder Tongue Family

* *Botrychium multifidum* (Gmel.) Trev.
Grape Fern
Rare near the outlet of Jenny Lake. (not seen in the last 40 years)

*Botrychium virginianum* (L.) Swartz
Grape Fern
Rare in the vicinity of Boy Scout Camp in Teton Canyon, west of western park boundary.

## Polypodiaceae Common Fern Family

* *Adiantum pedatum* L.
Northern Maidenhair Fern
Along streams on south facing slope 2 miles up Waterfalls Canyon, also Death Canyon.

* *Aspidotis densa* (Brack.) Lellinger
Indian Dream Fern
Rocks on the south facing slopes of Waterfalls Canyon, also slopes of Teewinot.

* *Asplenium trichomanes-ramosum* L. [*A. viride*]
Spleenwort
Rare in limestone at high elevations.

* *Athyrium alpestre* Clairv. var. *americanum* Butters [*A. distentifolium*]
Lady Fern
Small streams in North Fork of Cascade Canyon.

*Athyrium filix-femina* (L.) Roth
Lady Fern
Along streams draining into the valley lakes, also up to 8,000 ft. in the major canyons.

*Cheilanthes feei* Moore
Lip Fern
Rare, rocky sites at low to middle elevations.

*Cryptogramma acrostichoides* R. Br.
Rock Brake Fern
Rocky sites, lower portions of the major canyons.

*Cryptogramma stelleri* (Gmel.) Prantl
Rock Brake Fern
Rare, usually on wet limestone.

*Cystopteris fragilis* (L.) Bernh.
Bladder Fern
Rock crevices of the canyons.

* *Dryopteris expansa* (Presl) Fras.-Jenk. & Jermy [*D. assimilis*]
Shield Fern
Moist shaded woods, mouth of Cascade Creek.

* *Gymnocarpium dryopteris* (L.) Newm.
Oak Fern
Rare, moist wooded areas.

*Pellaea breweri* Eaton
Cliff Brake Fern
Rocky areas, usually limestone at mid- to high elevations.

*Pellaea glabella* Mett. ex Kuhn
Cliff Brake Fern
Rock crevices, north end of Blacktail Butte.

*Polystichum lonchitis* (L.) Roth
Holly Fern
Rocky crevices in major canyons and couloirs.

* *Polystichum scopulinum* (Eaton) Maxon
Mountain Holly Fern
Rare, the type locality is in upper Teton Canyon, Teton County.

*Pteridium aquilinum* (L.) Kuhn. var. *pubescens* Underw.
Bracken Fern
Around the western shores of the valley lakes. Uncooked fiddlenecks (unexpanded fronds) contain the enzyme thiaminase which causes thiamine deficiency.

*Woodsia oregana* Eaton
Woodsia Fern
Rocky and moist sites in the canyons.

*Woodsia scopulina* Eaton
Rocky Mountain Woodsia Fern
Among rocks, especially talus.

## DIVISION PINOPHYTA GYMNOSPERMS

### Cupressaceae Cypress Family

*Juniperus communis* L. var. *depressa* Pursh
Mountain Juniper
Shores of the valley lakes and extending into some canyons; circumboreal.

*Juniperus scopulorum* Sarg.
Rocky Mountain Juniper
Dry slopes of Blacktail Butte and in the Gros Ventre River bottoms.

### Pinaceae Pine Family

*Abies lasiocarpa* (Hook.) Nutt.
Subalpine Fir
Moraines and mountains up to tree line.

† *Larix occidentalis* Nutt.
Tamarack
Darby Canyon. Dorn (1988) says planted in Teton County.

*Picea engelmannii* Parry ex Engelm.
Engelmann Spruce
Western shores of the valley lakes and throughout the canyons, up to 9,000 ft.

*Picea pungens* Engelm.
Blue Spruce
Along the Snake River.

*Pinus albicaulis* Engelm.
Whitebark Pine
Mostly subalpine, above 8,000 ft.

*Pinus contorta* Dougl. ex Loud. var. *latifolia* Engelm. ex Wats.
Lodgepole Pine
Throughout the valley especially in potholes and on moraines.

*Pinus flexilis* James
Limber Pine
From windswept ridges in the valley to subalpine.

*Pseudotsuga menziesii* (Mirb.) Franco var. *glauca* (Beissn.) Franco
Douglas Fir
Moraines and in the canyons below 8,500 ft.

## DIVISION MAGNOLIOPHYTA FLOWERING PLANTS

## CLASS MAGNOLIOPSIDA DICOTS

### Aceraceae Maple Family

† *Acer ginnala* Maxim
Amur Maple
Cultivated at Black Rock Ranger Station; native of Asia. June.

*Acer glabrum* Torrey
Rocky Mountain Maple
Lower slopes of the mountains and into the canyons. May–June.

*Acer grandidentatum* Nutt.
Big Tooth Maple
Lower portions of the Snake River Canyon, very close to the southern boundary of Teton County. May–June.

### Amaranthaceae Amaranth Family

† *Amaranthus albus* L.
Tumble Pigweed
Widespread weed of disturbed sites; native of tropical America. June–August.

† *Amaranthus powellii* Wats.
Powell Amaranth
Weed of disturbed sites; native of tropical America. June–August.

† *Amaranthus retroflexus* L.
Redroot Pigweed
Weed of disturbed sites; native of tropical America. June–August.

### Anacardiaceae Cashew Family

*Toxicodendron rydbergii* (Small ex Rydb.) Greene
Poison Ivy
Rocky outcropping 150 yds. south of hot spring area, west side of Jackson Lake; contact with human skin causes painful skin irritation.

### Apiaceae Carrot Family

*Angelica arguta* Nutt.
Sharptooth Angelica
Stream banks and wet meadows of the valley. June–August.

*Angelica pinnata* Wats.
Pinnate Angelica
Moist places on the moraines of the valley. June–August.

*Angelica roseana* Henderson
Rose Angelica
Rocky areas in the mountains; Teton Canyon. July–August.

*Berula erecta* (Huds.) Cov. var. *incisa* (Torrey) Cronq.
Berula
In shallow water or wet areas. July–August.

*Bupleurum americanum* Coult. & Rose
American Thoroughwax
Alpine, east shore of Timberline Lake, 10,400 ft. July–August.

† *Carum carvi* L.
Caraway
Woods, meadows and roadsides; native of Europe. July–August.

† *Cicuta maculata* L. var. *angustifolia* Hook. [*C. douglasii*, misapplied]
Water Hemlock
Pond 1/4 mile south of Jackson Hole Airport, east side of highway, wet sites. Highly poisonous to man and livestock with cicutoxin concentrated in the base of the stem and roots; native of North America. June–August.

† *Conium maculatum* L.
Poison Hemlock
Roadside ditches and moist disturbed sites; native of Eurasia. This highly poisonous species is the "hemlock" of Socrates fame. June–August.

*Cymopterus longilobus* (Rydb.) Weber [*C. hendersonii*, misapplied]
Mountain Cymopterus
Major canyons above 9,000 ft. July–August.

*Cymopterus longipes* Wats.
Sprawling Cymopterus
Teton Pass; open sites, many soil types. May–June.

*Cymopterus terebinthinus* (Hook.) T. & G. var. *albiflorus* (T. & G.) Jones
Aromatic Cymopterus
Open rocky or sandy sites. May–June.

*Heracleum sphondylium* L. var. *lanatum* (Michx.) Dorn
Cow Parsnip
Moist areas throughout the valley and lower parts of canyons. June–August.

*Ligusticum filicinum* Wats.
Lovage
Woods and slopes of the valley and major canyons. July–August.

*Ligusticum porteri* Coult. & Rose
Porter Lovage
Woods, meadows and slopes, west of Flagg Ranch. June–August.

*Lomatium ambiguum* (Nutt.) Coult. & Rose
Lacy Lomatium
Roadside cuts north of Moose. June.

*Lomatium cous* (Wats.) Coult. & Rose
Cous Lomatium
Outwash east of Moose. June.

*Lomatium dissectum* (Nutt.) Math. & Const. var. *multifidum* (Nutt.) Math. & Const.
Giant Lomatium
Valley moraines and North Fork of Cascade Canyon. June–July.

*Lomatium foeniculaceum* (Nutt.) Coult. & Rose
Desert Parsley
Hills above the Gros Ventre River. May–June.

*Lomatium triternatum* (Pursh) Coult. & Rose
Ternate Lomatium
Area of the Oxbow Bend of the Snake River and Teton Canyon. June–July.

*Orogenia linearifolia* Wats.
Indian Potatoes; Turkey Peas
Sagebrush in the vicinity of Moose; blooming as soon as the ground is free of snow. April–May.

*Osmorhiza chilensis* H. & A.
Chilean Sweet Cicely
Shade of wooded areas of the valley. June–July.

*Osmorhiza depauperata* Phil.
Blunt Sweet Cicely
Wooded areas in the valley. June–July.

*Osmorhiza occidentalis* (Nutt. ex T. & G.) Torrey
Licorice-flavored Sweet Cicely
Aspen stands in the vicinity of the Oxbow Bend of the Snake River. June–July.

* *Perideridia bolanderi* (Gray) Nels. & Macbr.
Bolander Yampah
Hills, meadows and slopes near Moose. June–August.

*Perideridia montana* (Blank.) Dorn [*P. gairdneri* ssp. *borealis*]
Common Yampah
Open woods throughout the valley. Roots are edible. July–August.

*Sanicula marilandica* L.
Maryland Snakeroot
Open woods of the valley. June–July. (Reported by Alex Lasseigne)

*Sium suave* Walt.
Water Parsnip
Shallow pond north of Brinkerhoff Cabin. July–August.

*Zizia aptera* (Gray) Fern.
Zizia
Wet meadows, National Elk Refuge, 2 1/2 miles north of Jackson. July–August.

## Apocynaceae Dogbane Family

*Apocynum androsaemifolium* L.
Spreading Dogbane
Open woods and dry soils of moraines. All plant parts contain resins and glycosides capable of poisoning the hearts of mammals. June–August.

*Apocynum cannabinum* L.
Hemp Dogbane
Hills, slopes and sometimes disturbed sites in the valley. All plant parts contain resins and glycosides capable of poisoning mammals. June–August.

## Asclepiadaceae Milkweed Family

*Asclepias speciosa* Torrey
Showy Milkweed
Disturbed sites and stream banks, first seen along irrigation ditch west of Timbered Island. June–July.

## Asteraceae Sunflower Family

*Achillea millefolium* L. var. *lanulosa* (Nutt.) Piper
Milfoil Yarrow
Outwash, moraines and disturbed sites throughout the valley. June–August.

*Agoseris aurantiaca* (Hook.) Greene
Orange Dandelion
Meadows and open woods at moderate elevations in the mountains. June–August.

*Agoseris glauca* (Pursh) Raf.
Mountain Dandelion
Meadows at all elevations. June–August.

† *Ambrosia artemisiifolia* L.
Common Ragweed
Weed of disturbed sites; widespread in North America; causes hayfever. August–Sept.

*Anaphalis margaritacea* (L.) Benth. & Hook.
Pearly Everlasting
Open woods and slopes in the valley and lower elevations in the canyons. July–August.

*Antennaria alpina* (L.) Gaertn. var. *media* (Greene) Jeps.
Alpine Pussytoes
Subalpine cirques. July–August.

*Antennaria anaphaloides* Rydb.
Pearly Pussytoes
Slopes, meadows and open woods. June–August.

*Antennaria corymbosa* E. Nels.
Flattop Pussytoes
Sagebrush flats west of Moose. July–August.

*Antennaria dimorpha* (Nutt.) T. & G.
Low Pussytoes
Sagebrush flats of the valley. May–June.

*Antennaria luzuloides* T. & G.
Rush Pussytoes
Rocky flats of the valley and slopes of the Teton Range. June–August.

*Antennaria microphylla* Rydb.
Rose Pussytoes
Sagebrush flats and moraines of the valley. June–August.

*Antennaria parvifolia* Nutt.
Littleleaf Pussytoes
Dry terraces of the Pacific Creek area. June–July.

*Antennaria racemosa* Hook.
Raceme Pussytoes
North exposure of Signal Mt. and major canyons. July–August.

*Antennaria umbrinella* Rydb.
Umber Pussytoes
Outlet of Schoolroom Glacier Lake, 10,000 ft. July–August.

† *Anthemis cotula* L.
Mayweed; Chamomile
Weed of disturbed sites; native of Europe. June–August.

† *Anthemis tinctoria* L.
Yellow Chamomile
Cultivated outside the park and occasionally escaping; native of Europe. July–August.

† *Arctium minus* Bernh.
Common Burdock
Disturbed sites; native of Europe. June–July.

*Arnica chamissonis* Less. ssp. *foliosa* (Nutt.) Maguire
Chamisso Arnica
Moist sites in open forests, Snake River. June–August.

*Arnica cordifolia* Hook.
Heartleaf Arnica
Coniferous forests of the valley. Plants contain a volatile oil which is poisonous to children. June.

*Arnica latifolia* Bong.
Broadleaf Arnica
Major canyons 7,000–8,000 ft. July–August.

*Arnica longifolia* Eaton
Longfleaf Arnica
Moist sites of canyons and couloirs. July–August.

*Arnica mollis* Hook.
Hairy Arnica
Subalpine, in moist meadows. July–August.

*Arnica parryi* Gray
Parry Arnica
Open woods, meadows and slopes, Moose Ponds west of Moose. July–August.

*Arnica sororia* Greene
Twin Arnica
Slopes and meadows, near Moose. July.

† *Artemisia abrotanum* L.
Southernwood
Disturbed areas, remaining from previous cultivation. August.

*Artemisia arbuscula* Nutt. var. *arbuscula*
Low Sagebrush
West side of Snake River north of Moose. August–Sept.

*Artemisia biennis* Willd. var. *biennis*
Biennial Wormwood
Disturbed areas. August–Sept.

*Artemisia campestris* L. ssp. *borealis* (Pallas) Hall & Clem.
Field Wormwood
Rare in open places; circumboreal. August–Sept.

*Artemisia cana* Pursh var. *viscidula* Osterh.
Silver Sage
Along the Snake River. August–Sept.

*Artemisia dracunculus* L.
Tarragon
Dry exposed hillsides from the valley to subalpine. August–Sept.

*Artemisia frigida* Willd.
Fringed Sagebrush
Dry to mesic sites of the valley, often gravelly soil. August–Sept.

*Artemisia ludoviciana* Nutt.
var. *incompta* (Nutt.) Cronq.
Frequent in the canyons.
var. *latiloba* Nutt.
Infrequent in open places.
var. *ludoviciana*
Rare in the lowlands of the valley.
Prairie Sage
June–Sept.

*Artemisia michauxiana* Bess.
Michaux Wormwood
Rocky sites in the major canyons. July–August.

*Artemisia norvegica* Fries var. *saxatilis* (Bess.) Jeps.
Spruce Wormwood
Alpine or subalpine. July–August.

*Artemisia nova* A. Nels.
Black Sagebrush
Rocky hills of the valley. August–Sept.

*Artemisia scopulorum* Gray
Alpine Sagebrush
Subalpine and alpine; pine-spruce; talus. July–August.

*Artemisia tridentata* Nutt.
var. *tridentata*
var. *vaseyana* (Rydb.) Boivin
var. *wyomingensis* (Beetle & Young) Welsh
[*A. rothrockii*, misapplied]
Big Sagebrush
Common in the valley. August–Sept.

*Artemisia tripartita* Rydb. var. *tripartita*
Three-tipped Sagebrush
Well-drained areas of the valley. August–Sept.

*Aster alpigenus* (T. & G.) Gray var. *haydenii* (Porter) Cronq.
Alpine Aster
Shores of alpine lakes and some mountain summits. July–August.

*Aster ascendens* Lindl.
Long-leaved Aster
Hills, slopes and woods of the valley. August–Sept.

*Aster bracteolatus* Nutt. [*A. eatonii*]
Bracted Aster
Moist places in the valley, Taggart Creek. August–Sept.

*Aster campestris* Nutt.
Meadow Aster
Common along the Snake River. August–Sept.

*Aster conspicuus* Lindl.
Showy Aster
Common on the north side of Signal Mt. August–Sept.

*Aster engelmannii* (Eaton) Gray
Engelmann Aster
Shores of the valley lakes and in the major canyons up to 9,000 ft. July–Sept.

*Aster foliaceus* Lindl. ex DC.
Alpine Leafy Bract Aster
Common in the sagebrush and open forests of the valley. A variable species difficult to separate into varieties. July–Sept.

*Aster glaucodes* Blake
Blueleaf Aster
Talus slopes in major canyons. July–Sept.

*Aster hesperius* Gray
Ditchbank Aster
Shores, stream banks and meadows of the valley. August–Sept.

*Aster integrifolius* Nutt.
Thickstem Aster
Spruce-fir from the valley to the subalpine. August–Sept.

*Aster occidentalis* (Nutt.) T. & G.
Western Aster
Streamside forests and meadows. August–Sept.

*Aster perelegans* Nels. & Macbr.
Nuttall Aster
Outwash plain and extending up the mountain. August–Sept.

*Balsamorhiza macrophylla* Nutt.
Cutleaf Balsamroot
Along the Snake River and lower portion of Cottonwood Creek. June–July.

*Balsamorhiza sagittata* (Pursh) Nutt.
Arrowleaf Balsamroot
Sagebrush flats and on moraines of the valley. June.

*Bidens cernua* L.
Nodding Beggarticks
Margins of ponds and shallow lakes in the valley. July–August.

*Brickellia grandiflora* (Hook.) Nutt.
Tasselflower
Frequent on the lower portions of canyon trails such as Cascade Canyon. July–August.

† *Carduus nutans* L.
Musk Thistle
Weed of disturbed sites; native of Europe. July–Sept.

† *Centaurea maculosa* Lam.
Spotted Knapweed
Weed of disturbed sites; native of Europe; has invaded rapidly since 1980. August–Sept.

*Chaenactis alpina* (Gray) Jones
Alpine Chaenactis
Talus and boulder fields in the alpine. July–August.

*Chaenactis douglasii* (Hook.) H. & A. var. *montana* Jones
Douglas Chaenactis
Rocky slopes of the Snake River. June–August.

† *Chrysanthemum leucanthemum* L.
Oxeye Daisy
Disturbed moist sites; native of Eurasia. June–Sept.

*Chrysothamnus nauseosus* (Pallas ex Pursh) Britt.
Rubber Rabbitbrush
Hills and slopes on the east side of the valley. August–Sept.

*Chrysothamnus viscidiflorus* (Hook.) Nutt.
Low Rabbitbrush
Blacktail Pond overlook and 2 miles east of Elk Reservoir. August–Sept.

† *Cichorium intybus* L.
Chicory
Roadside weed; native of Eurasia. June–August.

† *Cirsium arvense* (L.) Scop.
Canada Thistle
Weed of disturbed sites; native of Europe. July–August.

*Cirsium eatonii* (Gray) Robins.
Eaton Thistle
Mountain slopes and talus in the alpine. July–August.

*Cirsium scariosum* Nutt.
Elk Thistle
Wet meadows throughout the valley. July–August.

*Cirsium subniveum* Rydb.
Snow Thistle
Slopes and flats at lower end of valley. July–August.

*Cirsium tweedyi* (Rydb.) Petr.
Tweedy Thistle
Moraines of alpine glaciers.

† *Cirsium vulgare* (Savi) Tenore
Bull Thistle
Weed of disturbed sites; native of Eurasia. July–August.

† *Conyza canadensis* (L.) Cronq.
Canada Fleabane
Weed of disturbed sites; native of North America. August–Sept.

*Crepis acuminata* Nutt.
Tapertip Hawksbeard
Streambeds and open forests of the valley. June–July.

*Crepis atrabarba* Heller
Slender Hawksbeard
Sagebrush and coniferous woods, north slope of Signal Mt. June.

*Crepis modocensis* Greene
Modoc Hawksbeard
Moraines and outwash plains of the valley. June.

*Crepis runcinata* (James) T. & G. var. *glauca* (Nutt.) Boivin
Dandelion Hawksbeard
Moist sites in the valley. June–August.

† *Crepis tectorum* L.
Rooftop Hawksbeard
Stream banks and disturbed areas; native of Europe. June–July.

*Erigeron acris* L. var. *asteroides* (Andrz. ex Bess.) DC.
Bitter Fleabane
Open, rocky places in the mountains. June–July.

*Erigeron compositus* Pursh var. *discoideus* Gray
Fernleaf Fleabane
Subalpine to alpine; rocky slopes. July–August.

*Erigeron corymbosus* Nutt.
Mountain Daisy
Sagebrush and rabbitbrush. July.

*Erigeron divergens* T. & G.
Spreading Fleabane
Sagebrush flats and along the Snake River. June–August.

*Erigeron eatonii* Gray
Eaton Fleabane
Rocky sites of the canyons and valley. June–August.

*Erigeron flagellaris* Gray
Trailing Fleabane
Sagebrush to spruce-fir. June–August.

*Erigeron glabellus* Nutt.
Smooth Fleabane
Wet meadows along the Snake River. July–August.

*Erigeron leiomerus* Gray
Glaber Fleabane
Subalpine to alpine; crevices and cliffs. July–August.

*Erigeron lonchophyllus* Hook.
Spearleaf Fleabane
Along streams north of Jackson Lake Dam. June–August.

*Erigeron ochroleucus* Nutt.
Cream-colored Fleabane
Sagebrush areas along the Snake River. June–August.

*Erigeron peregrinus* (Banks ex Pursh) Greene ssp. *callianthemus* (Greene) Cronq.
Peregrine Fleabane
Midmontane to subalpine; mesic sites. June–August.

*Erigeron pumilus* Nutt.
Low Fleabane
Sagebrush flats throughout the valley. June.

*Erigeron rydbergii* Cronq.
Rydberg Fleabane
Subalpine and alpine, Sheep Mt. 9,800 ft. August.

*Erigeron simplex* Greene
Oneflowered Fleabane
Subalpine and alpine. July–August.

*Erigeron speciosus* (Lindl.) DC.
Oregon Fleabane
Open woods, meadows and slopes throughout the valley; a highly variable species. July–August.

*Erigeron ursinus* Eaton
Bear River Fleabane
Ledges and cliffs in the subalpine and alpine. July–August.

*Eriophyllum lanatum* (Pursh) Forbes var. *integrifolium* (Hook.) Smiley
Pursh Woollyleaf
Sagebrush flats and on the moraines of the valley; a highly variable species. June–August.

*Gaillardia aristata* Pursh
Common Blanket Flower
Valley, Kelly Post Office (R. Lichvar 1278). July–August.

*Gnaphalium chilense* Spreng.
Cottonbatting Cudweed
Moist places, Snake River, 20 miles south of Jackson. July–August.

*Gnaphalium microcephalum* Nutt.
White Cudweed
Rare in Snowshoe Canyon. July–August.

*Gnaphalium palustre* Nutt.
Lowland Cudweed
Common along the Snake River. July–August.

*Gnaphalium viscosum* H.B.K.
Sticky Cudweed
Open woods and slopes, Cascade Canyon. July–August.

*Grindelia squarrosa* (Pursh) Dunal var. *squarrosa*
Curlycup Gumweed
Common along roads and disturbed sites. July–August.

*Gutierrezia sarothrae* (Pursh) Brit. & Rusby
Broom Snakeweed
Dry, open sites on the National Elk Refuge. August–Sept.

*Haplopappus acaulis* (Nutt.) Gray
Stemless Goldenweed
Ridges and slopes, west side of Blacktail Butte. August–Sept.

*Haplopappus lyallii* Gray
Lyall Goldenweed
Alpine or subalpine, Togwotee Pass. July–August.

*Haplopappus macronema* Gray var. *linearis* (Rydb.) Dorn
Whitestem Goldenweed
Rocky sites in the subalpine. July–August.

*Haplopappus suffruticosus* (Nutt.) Gray
Shrubby Goldenweed
Alpine or subalpine, major canyons and Alaska Basin. July–August.

*Haplopappus uniflorus* (Hook.) T. & G.
One-flowered Goldenweed
Dry, open sites, east side of park. July–August.

*Helenium hoopesii* Gray
Orange Sneezeweed
Bridger-Teton National Forest east of Blackrock Ranger Station. July–August.

*Helianthella quinquenervis* (Hook.) Gray
Five-veined Helianthella
Morainal soils throughout the west side of the valley. July–August.

*Helianthella uniflora* (Nutt.) T. & G.
One-flowered Helianthella
Morainal soils of the valley and extending into the canyons. June–August.

*Helianthus annuus* L.
Common Sunflower
Widespread in disturbed sites; native of North America. July–August.

*Helianthus nuttallii* T. & G.
Nuttall Sunflower
Moist sites. July–August.

*Helianthus petiolaris* Nutt.
Prairie Sunflower
Widespread as a weed; native of North America. July–August.

*Heterotheca depressa* (Rydb.) Dorn
Golden Aster
Sandy or gravelly sites along the Snake River. July–August.

*Heterotheca horrida* (Rydb.) Harms [*Chrysopsis horrida*]
Horrible Golden Aster
Gros Ventre River; one collection in 1894. August.

*Heterotheca villosa* (Pursh) Shinners
Hairy Golden Aster
Open rocky hills and slopes of the valley; a highly variable species. July–Sept.

*Hieracium albiflorum* Hook.
White Hawkweed
Shade of lodgepole pine forest. July–Sept.

*Hieracium cynoglossoides* Arv.-Touv.
Houndstongue Hawkweed
Woods of the valley. July–August.

*Hieracium gracile* Hook.
Slender Hawkweed
Subalpine meadows and woods. July–August.

* *Hieracium scouleri* Hook.
Woolly Hawkweed
Open woods and sagebrush flats of the valley. July–August.

*Hymenoxys grandiflora* (T. & G. ex Gray) Parker
Rydberg Sunflower
Subalpine and alpine talus slopes. July–August.

† *Iva xanthifolia* Nutt.
Marshelder
Moist disturbed sites in the valley; native weed of middle North America. July–Sept.

† *Lactuca serriola* L.
Prickly Lettuce
Weed of disturbed sites; native of Europe. July–Sept.

*Lactuca tatarica* (L.) Mey. [*L. pulchella*]
Chicory Lettuce
Weed of disturbed sites; circumboreal. July–August.

*Machaeranthera canescens* (Pursh) Gray var. *canescens*; var. *monticola* Dorn
Hoary Aster
Sagebrush flats of the valley. July–Sept.

*Madia glomerata* Hook.
Cluster Tarweed
Disturbed sites of the valley; native weed. July–August.

† *Matricaria matricarioides* (Less.) Porter
Pineapple Weed
Weed of disturbed sites; native of North America. June–August.

*Microseris nutans* (Hook.) Schultz-Bip.
Nodding Microseris
Lupine meadows south of Jenny Lake. June–July.

† *Ratibida columnifera* (Nutt.) Wooton & Standley
Prairie Coneflower
Occasionally seen along roadsides; appears to be adventive from the Great Plains. July–August.

*Rudbeckia occidentalis* Nutt.
Western Coneflower
Moist places in the valley and canyons; in some areas this species indicates overgrazing. August–Sept.

*Senecio amplectens* Gray var. *holmii* (Greene) Harrington
Alpine Groundsel
Limestone in Moose Basin and South Teton Canyon. July–August.

*Senecio canus* Hook.
Gray Groundsel
Windswept ridges of the alpine. July–August.

*Senecio crassulus* Gray
Thickleaf Groundsel
Meadows and woods of the valley and in some of the major canyons. June–August.

*Senecio debilis* Nutt.
Weak Butterweed
Moist meadows, Gros Ventre Mts. July–August.

*Senecio dimorphophyllus* Greene var. *paysonii* Barkley
Main couloir on the east face of Symmetry Spire, 9,000 ft. July–August.

*Senecio fremontii* T. & G. var. *fremontii*
Fremont Groundsel
Rocky places in the alpine; North Fork of Cascade Canyon. July–August.

*Senecio hydrophilus* Nutt.
Water Groundsel
Ponds east of Colter Bay Village. July–August.

*Senecio integerrimus* Nutt.
Columbia Groundsel
Sagebrush flats throughout the valley; an indicator of overgrazing. May–June.

*Senecio pauperculus* Michx.
Balsam Groundsel
Moist meadows in the valley. July–August.

*Senecio serra* Hook.
Saw Groundsel
Moist sites along Cottonwood Creek. July–Sept.

*Senecio sphaerocephalus* Greene
Roundhead Groundsel
Disturbed sites in Willow Flats north of Jackson Lake Dam. July–Sept.

*Senecio streptanthifolius* Greene
Manyface Groundsel
Wooded area and meadow, Oxbow Bend area. June–August.

*Senecio triangularis* Hook.
Arrowleaf Groundsel
Canyons to subalpine, moist shady sites. July–Sept.

*Senecio werneriifolius* Gray var. *alpinus* (Gray) Dorn
Rock Butterweed
Alpine and subalpine; July–August.

*Solidago canadensis* L. var. *salebrosa* (Piper) Jones
Canada Goldenrod
Along the Snake River in disturbed sites. July–Sept.

*Solidago gigantea* Ait.
Smooth Goldenrod
Open moist places, Double Diamond Ranch (Climber Ranch). July–August.

*Solidago glutinosa* Nutt. [*S. spathulata*]
Coast Goldenrod
Meadows east and north of Jackson Lake Dam. July–Sept.

*Solidago missouriensis* Nutt. var. *missouriensis*
Missouri Goldenrod
Streamside forest along the Snake River. July–Sept.

*Solidago mollis* Bartl.
Velvety Goldenrod
Open woods, Jenny Lake area. July–Sept.

*Solidago multiradiata* Ait. var. *scopulorum* Gray
Low Goldenrod
Along the Snake River and lower portions of the canyons. July–August.

*Solidago nana* Nutt.
Dwarf Goldenrod
Hills, slopes and open woods of the valley. July–Sept.

† *Sonchus oleraceus* L.
Common Sowthistle
Weed of disturbed sites; native of Europe. June–August.

*Stephanomeria tenuifolia* (Raf.) Hall
Narrow-leaf Skeletonweed
Wet sand and rocky bars of the Snake River. July–August.

† *Tanacetum vulgare* L.
Common Tansy
Along the Snake River below the Jackson Lake Dam, remaining from previous cultivation; native of Europe. Humans have been poisoned by drinking tea made from leaves and flowers. July–Sept.

† *Taraxacum laevigatum* (Willd.) DC.
Red-fruited Dandelion
Weed of disturbed sites throughout the valley; native of Europe. June–August.

† *Taraxacum officinale* Weber
Common Dandelion
Weed of disturbed sites; native of Eurasia. May–August.

*Taraxacum scopulorum* (Gray) Rydb. [*T. lyratum*, misapplied]
Alpine Dandelion
Rocky slopes and talus in the alpine. July–August.

*Tetradymia canescens* DC.
Gray Horsebrush
Dry gravelly sites in the valley. June–Sept.

*Townsendia florifer* (Hook.) Gray
Showy Townsendia
Mountain slopes. June–August.

*Townsendia hookeri* Beaman
Hooker Townsendia
Barren sites on east side of valley. July–August.

*Townsendia leptotes* (Gray) Osterh.
Common Townsendia
Slopes and ridges in the subalpine. June–July.

*Townsendia montana* Jones [*T. alpigena*]
Mountain Townsendia
Slopes and ridges in the subalpine and alpine. July–August.

*Townsendia parryi* Eaton
Parry Townsendia
Open areas in the mountains. July–August.

† *Tragopogon dubius* Scop.
Yellow Salsify; Goatsbeard
Weed of disturbed sites; native of Europe. June–August.

† *Tragopogon porrifolius* L.
Purple Salsify; Goatsbeard
Weed of waste places and roadsides; native of Europe. June–August.

† *Tragopogon pratensis* L.
Meadow Salsify; Goatsbeard
Weed of moist disturbed sites; native of Europe. June–August.

*Viguiera multiflora* (Nutt.) Blake
Showy Goldeneye
Hills and slopes in open woods of the valley. July–Sept.

*Wyethia amplexicaulis* (Nutt.) Nutt.
Mulesears
West of Cattleman's Bridge on the Snake River. June–July.

*Wyethia helianthoides* Nutt.
White-head Mulesears
Moist meadows, Highway 89, 4 miles north of Colter Bay. June.

† *Xanthium strumarium* L. var. *canadense* (Miller) T. & G.
Cocklebur
Weed of disturbed sites; weed of doubtful origin; reported as poisonous to livestock. August–Sept.

## Berberidaceae Barberry Family

*Mahonia repens* (Lindl.) G. Don
Common Oregon Grape
Woods around the valley lakes. May–June.

## Betulaceae Birch Family

*Alnus incana* (L.) Moench var. *occidentalis* (Dippel) Hitchc.
Mountain Alder
Moist places around the shores of the valley lakes and along the Snake River. May–June.

*Betula glandulosa* Michx.
Glandular Birch
Meadows and wet places of the valley and some canyons. June–July.

*Betula occidentalis* Hook.
Water Birch
Rare on the south end of Blacktail Butte. May–June.

## Boraginaceae Borage Family

*Amsinckia menziesii* (Lehm.) Nels. & Macbr.
Menzies Fiddleneck
Hills and disturbed sites on the east side of the valley. May–July.

† *Asperugo procumbens* L.
Catchweed
Weed of disturbed sites; native of Eurasia. May–June.

* *Cryptantha affinis* (Gray) Greene
Slender Cryptantha
Woods and slopes throughout the valley. June–July.

*Cryptantha ambigua* (Gray) Greene
Wilkes Cryptantha
Dry open sites, near Moose. June–July.

*Cryptantha torreyana* (Gray) Greene
Torrey Cryptantha
Dry, open sites in the valley. June–August.

*Cryptantha watsonii* (Gray) Greene
Watson Cryptantha
Disturbed sites at Colter Bay Village. June–July.

† *Cynoglossum officinale* L.
Hounds Tongue
Weed of disturbed sites; native of Europe. June–July.

*Eritrichium nanum* (Vill.) Schrad. ex Gaudin var. *elongatum* (Rydb.) Cronq.
Alpine Forget-me-not
Ridges and slopes in the alpine above 10,000 ft.; official flower of Grand Teton National Park. July–August.

*Hackelia floribunda* (Lehm.) Johnston
Biennial Forget-me-not
Meadows, woods and disturbed sites up to 9,000 ft. June–August.

*Hackelia micrantha* (Eastw.) Gentry
Meadow Forget-me-not
Meadows to stream banks in the valley. June–August.

*Hackelia patens* (Nutt.) Johnston
Spotted Forget-me-not
Open woods and slopes of the valley. June–August.

*Lappula redowskii* (Hornem.) Greene
Desert Stickseed
Disturbed sites along the Snake River. May–June.

*Lithospermum ruderale* Dougl. ex Lehm.
Western Stoneseed
Along the Snake River and extending out into the sagebrush flats. May–June.

*Mertensia ciliata* (James ex Torrey) G. Don
Streamside Bluebell
Stream banks and wet meadows up to 9,500 ft. June–August.

*Mertensia oblongifolia* (Nutt.) G. Don
Sagebrush Bluebell
Sagebrush flats throughout the valley. June–July.

*Mertensia viridis* (A. Nels.) A. Nels.
Mountain Bluebell
Subalpine and alpine, open rocky sites, southern part of Teton County. July–August.

*Myosotis alpestris* Schmidt
Wood Forget-me-not
Meadows and slopes in the mountains; subalpine and alpine. July–August.

† *Myosotis scorpioides* L.
True Forget-me-not
Snake River bottom south of Moose; native of Europe. June–July.

*Plagiobothrys scouleri* (H. & A.) Johnston var. *hispidulus* (Greene) Dorn
Meadow Plagiobothrys
Moist, poorly drained soil in the valley. June–August.

## Brassicaceae Mustard Family

† *Alyssum alyssoides* (L.) L.
Yellow Alyssum
Weed of disturbed sites; native of Europe. May–June.

† *Alyssum desertorum* Stapf
Desert Alyssum
Weed of disturbed sites; native of Eurasia. May–June.

*Arabis drummondii* Gray
Drummond Rockcress
Moraines and outwash plains of the valley. June–August.

*Arabis glabra* (L.) Bernh.
Tower Mustard
Sagebrush to woods of the valley. June–July.

*Arabis hirsuta* (L.) Scop.
Hairy Rockcress
Meadows east of the Oxbow Bend of the Snake River; circumboreal. June–August.

*Arabis holboellii* Hornem.
Holboell Rockcress
Sagebrush flats and moraines of the valley; a variable species with three varieties listed for our area. May–June.

*Arabis lemmonii* Wats.
Lemmon Rockcress
Alpine moraines. July–August.

*Arabis lyallii* Wats.
Lyall Rockcress
Subalpine and alpine areas on rocky ridges. July–August.

*Arabis microphylla* Nutt. var. *microphylla*
Small-leaf Rockcress
Woods and slopes in the subalpine, Teton Canyon. June–July.

*Arabis nuttallii* Robins.
Nuttall Rockcress
Meadows east of the Oxbow Bend of the Snake River. June–July.

*Barbarea orthoceras* Ledeb.
Erectpod Winter Cress
Meadows and open woods of the valley; circumboreal. June–July.

† *Barbarea vulgaris* R. Br.
Yellowrocket Winter Cress
Weed of disturbed sites; native of Eurasia. June–July.

† *Berteroa incana* (L.) DC.
Hoary Alyssum
Weed of roadsides; native of Europe. June–August.

† *Brassica juncea* (L.) Czern.
Indian Mustard
Weed of disturbed sites; native of Asia. June–August.

† *Brassica kaber* (DC.) Wheeler
Wild Mustard
Weed of disturbed sites; native of Europe. June–July.

† *Capsella bursa-pastoris* (L.) Medic.
Shepherds Purse
Weed of disturbed sites; native of Europe. May–July.

*Cardamine breweri* Wats.
Brewer Bittercress
Wet areas along the Snake River. June–August.

† *Cardamine oligosperma* Nutt.
Nuttall Bittercress
Weed of moist disturbed sites; apparently a native of Eurasia. June–July.

† *Chorispora tenella* (Pallas) DC.
Musk Mustard
Weed of disturbed sites; native of Asia. May–June.

*Descurainia californica* (Gray) Schulz
California Tansy Mustard
Slopes and open woods, Snow King Mt. June–July.

*Descurainia incana* (Bernh. ex Fisch. & Meyer) Dorn [*D. richardsonii*]
Tansy Mustard
Hills, slopes and meadows of the valley; four varieties have been described. June–July.

*Descurainia pinnata* (Walt.) Britt.
Pinnate Tansy Mustard
Hills, slopes and disturbed sites; this taxon is composed of numerous intergrading variants. May–August.

† *Descurainia sophia* (L.) Webb ex Prantl
Flixweed Tansy Mustard
Weed of disturbed sites; native of Eurasia. May–August.

*Draba aurea* Vahl ex Hornem.
Golden Draba
Woods, slopes and meadows in the mountains. June–August.

* *Draba borealis* DC.
Boreal Draba
Rocky slopes and meadows. June–July.

* *Draba crassa* Rydb.
Thick Draba
Alpine zone on cliffs above 10,000 ft. July–August.

*Draba crassifolia* Grah.
Hairy Draba
Subalpine to alpine on talus slopes. June–August.

* *Draba globosa* Payson [*D. apiculata*]
Pointed Draba
East shore of Timberline Lake, 10,400 ft. July–August.

*Draba lonchocarpa* Rydb.
Lancefruit Draba
Subalpine and alpine on rocky sites. July–August.

*Draba nemorosa* L.
Woods Draba
Moraines of the valley. May–June.

*Draba oligosperma* Hook.
Doublecomb Draba
Rocky slopes and exposed ridges in the alpine. July–August.

*Draba praealta* Greene
Tall Whitlow-wort
Rocky slopes and woods, Hoback Canyon. June.

*Draba ventosa* Gray
Wind Draba
Ridges and slopes in alpine zone; summit of Mt. Moran. July–August.

*Erysimum asperum* (Nutt.) DC.
Rough Wallflower
Infrequent in the lower portions of the canyons. June–July.

*Erysimum cheiranthoides* L.
Treacle Mustard
Moist sites in meadows around Signal Mt. June–August.

*Erysimum inconspicuum* (Wats.) MacM.
Lesser Wallflower
Dry open areas; only one specimen dated 1894 (A. Nelson—916). August.

† *Hesperis matronalis* L.
Dames Violet; Sweet Rocket
Cultivated and escaping along irrigation ditches; native of Europe; Snow King Mt. June–August.

† *Isatis tinctoria* L.
Dyers Woad
Weed of disturbed sites; native of Eurasia; first seen at Colter Bay Visitor Center about 1975. June–July.

† *Lepidium campestre* (L.) R. Br.
Fieldcress
Weed of disturbed sites; native of Europe. May–July.

*Lepidium densiflorum* Schrad. var. *pubecarpum* (A. Nels.) Thell.
Densecress
Disturbed areas in the valley. May–July.

† *Lepidium latifolium* L.
Perennial Pepperweed
Weed of disturbed sites; native of Eurasia. June–August.

† *Lepidium perfoliatum* L.
Clasping Pepperweed
Weed of disturbed sites; native of Europe. May–June.

*Lesquerella alpina* (Nutt.) Wats. var. *alpina*
Alpine Bladderpod
Hills and slopes. May–July.

* *Lesquerella carinata* Rollins
Keeled Bladderpod
Hills and slopes of the Gros Ventre River drainage. June–July.

* *Lesquerella paysonii* Rollins
Payson Bladderpod
Rocky slopes and gravel bars, mouth of Pacific Creek; Marriott lists this species as a narrow endemic. June–July.

† *Nasturtium officinale* R. Br.
Water Cress
Delta at the mouth of Cascade Creek, west shore of Jenny Lake; native of Eurasia. June–August.

*Physaria acutifolia* Rydb. [*P. australis*]
Common Twinpod
Rocky slopes and open woods. June–July.

*Physaria didymocarpa* (Hook.) Gray
Twinpod
Open rocky slopes along Gros Ventre River. June–July.

*Rorippa curvipes* Greene var. *curvipes*; var. *alpina* (Wats.) Stuckey
Curved Yellowcress
Wet sites in the valley, shore of Jackson Lake. June–August.

* *Rorippa curvisiliqua* (Hook.) Bessey ex Britt.
var. *curvisiliqua*; var. *orientalis* Stuckey
Western Yellowcress
Moist sites in the valley. June–August.

*Rorippa palustris* (L.) Besser ssp. *fernaldiana* (B. & A.) Jonsell
Marsh Yellowcress
Wet areas along the Snake River. June–August.

† *Sisymbrium altissimum* L.
Tumbling Mustard
Weed of disturbed sites; native of Europe. June–July.

*Smelowskia calycina* (Steph. ex Willd.) Meyer var. *americana* (R. & H.) Drury & Rollins
Alpine Smelowskia
Exposed rocky slopes in the alpine. July–August.

*Subularia aquatica* L.
Awlwort
Usually submerged in ponds, near Togwotee Pass. June–August.

*Thelypodium paniculatum* A. Nels.
Panicled Thelypody
Hills and meadows, only one specimen—1894 (A. Nelson—922). June–August.

† *Thlaspi arvense* L.
Field Pennycress
Weed of disturbed sites; native of Europe. May–June.

*Thlaspi montanum* L. var. *montanum*
Wild Candytuft
Moist to dry rocky slopes, 6,000 to 8,500 ft. June–August.

*Thlaspi parviflorum* A. Nels.
Smallflowered Pennycress
Meadows and stream banks of the valley. June–August.

## Cactaceae Cactus Family

*Opuntia fragilis* (Nutt.) Haw.
Brittle Pricklypear
Gravel banks on the Snake River in area of Deadman's Bar, also at Kelly Warm Spring. June–July.

*Opuntia polycantha* Haw.
Plains Pricklypear
West of Highway 89, opposite Elk Refuge and 2–3 miles north of Jackson. June–July.

## Callitrichaceae Water Starwort Family

*Callitriche hermaphroditica* L.
Autumn Water Starwort
Aquatic in ponds and lakes; circumboreal. June–Sept.

*Callitriche palustris* L.
Water Starwort
Pools and adjacent mud along the Snake River. July–Sept.

## Campanulaceae Bellflower Family

*Campanula rotundifolia* L.
Scotch Bluebell
Moraines, slopes and woods of the valley and canyons. June–August.

*Heterocodon rariflorum* Nutt.
Heterocodon
Moist places in mountains; reported in Teton Canyon. July–August.

*Porterella carnosula* (H. & A.) Torrey
Fleshy Porterella
Locally common at the margins of drying ponds north of Leigh Lake. July–August.

## Cannabaceae Hemp Family

† *Humulus lupulus* L. var. *neomexicanus* Nels. & Cock.
Common Hops
Persisting from cultivation at some of the old valley ranches. Hops secrete lupulin which is used to flavor beer. July–August.

## Caprifoliaceae Honeysuckle Family

*Linnaea borealis* L. var. *longiflora* Torrey
Longtube Twinflower
Shade of Lodgepole Pine; slopes of Signal Mt. July–August.

*Lonicera involucrata* Banks ex Spreng.
Bearberry Honeysuckle
Coniferous woods around the shores of the valley lakes. June–July.

† *Lonicera tatarica* L.
Tatarian Honeysuckle
Cultivated shrub at some of the old valley ranches; native of Eurasia. May–June.

*Lonicera utahensis* Wats.
Utah Honeysuckle
Open coniferous woods of the valley and canyons. June–July.

*Sambucus racemosa* L.
ssp. *pubens* (Michx.) House
var. *melanocarpa* (Gray) McMinn; var. *microbotrys* (Rydb.) Kearney & Peebles
Red Elderberry
Shores of the valley lakes and extending into the canyons. Stems, leaves and unripe berries contain a poisonous alkaloid. June–July.

*Symphoricarpos albus* (L.) Blake var. *laevigatus* (Fern.) Blake
White Snowberry
Moraines and sagebrush flats of the valley. June–July.

*Symphoricarpos occidentalis* Hook.
Wolfberry
Woods and riparian habitat of the valley. June–August.

*Symphoricarpos oreophilus* Gray var. *utahensis* (Rydb.) A. Nels.
Mountain Snowberry
Woods, meadows and slopes of the valley. June–August.

## Caryophyllaceae Pink Family

† *Agrostema githago* L.
Corn Cockle
Weed of disturbed sites; native of Eurasia; seeds are reputed to be poisonous to animals. June–July.

*Arenaria congesta* Nutt.
Congested Sandwort
Sagebrush, open woods and dry ridges from the valley to alpine. June–August.

*Arenaria lateriflora* L. [*Moehringia lateriflora*]
Bluntleaf Sandwort
Moist woods and meadows of the valley. June–August.

*Arenaria nuttallii* Pax
Nuttall Sandwort
Rocky soil from sagebrush flats to alpine. June–August.

*Arenaria obtusiloba* (Rydb.) Fern.
Arctic Sandwort
Subalpine to alpine, talus. July–August.

*Arenaria stricta* Michx. var. *dawsonensis* (Britt.) Scoggan
Slender Sandwort
Moist areas, Hoback Canyon. June–August.

*Cerastium arvense* L.
Field Chickweed
Dry to moist meadows, valley to alpine; a perplexing polymorphic species. June–August.

*Cerastium beeringianum* Cham. & Schlecht.
Bering Chickweed
Moist meadows in subalpine and alpine. June–August.

† *Cerastium fontanum* Baumg.
Mouse-ear Chickweed
Lawn weed; native of Europe. June–July.

† *Dianthus armeria* L.
Grass Pink
Disturbed sites; native of Europe. June–August.

† *Gypsophila paniculata* L.
Babys Breath
Escape from cultivation; native of Eurasia. June–August.

*Sagina saginoides* (L.) Karsten
Arctic Pearlwort
Moist sites, midmontane to alpine; circumboreal. June–August.

† *Saponaria officinalis* L.
Bouncing Bet
Escaped ornamental, behind Park Headquarters building; native of Europe. July–August.

*Silene acaulis* (L.) Jacq. var. *subacaulescens* (Wms.) Fern. & St. John
Moss Campion
Exposed ridges in the alpine; circumboreal. July–August.

† *Silene dichotoma* Ehrh.
Forked Catchfly
Weed of disturbed sites; a native of Eurasia. June–July.

† *Silene latifolia* Poir. [*Lychnis alba*]
White Campion
Weed of disturbed sites; native of Europe. July–August.

*Silene menziesii* Hook.
Menzies Silene
Dry to mesic sites in woods and slopes in the valley. June–August.

† *Silene noctiflora* L.
Night-flowering Catchfly
Weed of disturbed sites; native of Europe. June–August.

*Silene oregana* Wats.
Oregon Silene
Open woods and subalpine slopes in Teton Canyon. June–August.

*Silene parryi* (Wats.) Hitchc. & Maguire
Parry Campion
Meadows and slopes in the canyons above 9,500 ft. July–August.

† *Spergularia rubra* (L.) J. & K. Presl
Red Sandspurry
Weed of disturbed sites; native of Europe. July–Sept.

*Stellaria calycantha* (Ledeb.) Bong.
Northern Starwort
Woods and meadows of the canyons; circumboreal. June–August.

*Stellaria crassifolia* Ehrh.
Thickleaved Starwort
Moist areas. June–August.

*Stellaria crispa* Cham. & Schlecht.
Crisped Starwort
Moist areas, below Phelps Lake. June–August.

*Stellaria jamesiana* Torrey
Sticky Starwort
Woods, meadows and slopes in the valley, Hoback Canyon. June–August.

*Stellaria longifolia* Muhl. ex Willd.
Longleaved Starwort
Meadows and stream banks along the Snake River. June–August.

*Stellaria longipes* Goldie
Longstalk Starwort
Oxbow Bend of Snake River. June–August.

*Stellaria obtusa* Engelm.
Blunt Starwort
Stream banks and meadows of the valley. July–August.

*Stellaria umbellata* Turcz.
Umbrella Starwort
Woods, slopes and meadows in the subalpine. June–August.

† *Vaccaria pyramidata* Medic.
Pyramid Soapwort
Weed of disturbed sites; native of Europe. June–August.

## Celastraceae Stafftree Family

*Paxistima myrsinites* (Pursh) Raf.
Mountain Lover
Moraines, shores of the valley lakes and extending up to 9,000 ft. in the mountains. June.

## Ceratophyllaceae Hornwort Family

*Ceratophyllum demersum* L.
Common Hornwort
Lakes, ponds and slow moving streams in the valley. June–Sept.

## Chenopodiaceae Goosefoot Family

*Atriplex patula* L.
Fat-hen Saltbush
Weed of disturbed sites; a circumboreal variable complex. July–Sept.

† *Chenopodium album* L.
Lambsquarter; Pigweed
Weed of disturbed sites; native of Eurasia. Greens and seeds contain toxic quantities of nitrate and oxalic acid. June–Sept.

*Chenopodium atrovirens* Rydb.
Mountain Goosefoot
Disturbed sites in the valley. July–Sept.

*Chenopodium capitatum* (L.) Asch.
Strawberry Spinach
Moist places in the mountains. June–Sept.

*Chenopodium glaucum* L. var. *salinum* (Standl.) Boivin
Oakleaf Goosefoot
Moist areas, upper Gros Ventre River. July–Sept.

*Chenopodium leptophyllum* (Moq.) Nutt. ex Wats.
Narrowleaf Goosefoot
Disturbed sites, valley to canyons. July–Sept.

*Chenopodium rubrum* L. var. *glomeratum* Wallr. [*C. chenopodioides*]
Red Goosefoot
Moist sites in or near the mountains; appears to be circumboreal. August–Sept.

*Krascheninnikovia lanata* (Pursh) Meeuse & Smit [*Ceratoides lanata*]
Winterfat
Saline areas east of Elk Ranch Reservoir. June–Sept.

† *Kochia scoparia* (L.) Schrad.
Summer Cypress
Weed of disturbed sites; native of Europe. August–Sept.

*Monolepis nuttalliana* (Schultes) Greene
Nuttall Povertyweed
Dry to moist disturbed sites in the valley. June–Sept.

† *Salsola australis* R. Br. [*S. kali*]
Russian Thistle; Tumbleweed
Weed of disturbed sites; native of Asia. July–Sept.

*Sarcobatus vermiculatus* (Hook.) Torrey
Black Greasewood
Moist alkaline areas east of Elk Ranch Reservoir. June–July.

*Suaeda calceoliformis* (Hook.) Moq. [*S. depressa*]
Slender Seepweed
Alkaline areas, 2 miles east of Elk Ranch Reservoir. August–Sept.

## Clusiaceae St. Johnwort Family

*Hypericum formosum* H. B. K. var. *scouleri* (Hook.) Coult.
Western St. Johnwort
Moist open sites in the canyons above 8,000 ft. July–August.

† *Hypericum perforatum* L.
Common St. Johnwort
Weed of disturbed sites; native of Europe; poisonous to livestock because of hypericin which causes animals to be allergic to sunlight. July–Sept.

## Convolvulaceae Morning-glory Family

† *Convolvulus arvensis* L.
Field Bindweed
Weed of disturbed sites; native of Europe. July–Sept.

## Cornaceae Dogwood Family

*Cornus sericea* L. [*C. stolonifera*]
Redoiser Dogwood
Streamsides and moist woods in the valley. June–July.

## Crassulaceae Stonecrop Family

*Sedum debile* Wats.
Weakstemmed Stonecrop
Rocky ledges along the lower slopes of the mountains. June–August.

*Sedum integrifolium* (Raf.) A. Nels. [*S. rosea,* misapplied]
Roseroot Stonecrop
Moist meadows and stream banks, subalpine to alpine. July–Sept.

*Sedum lanceolatum* Torr. [*S. stenopetalum* Pursh, misapplied]
Lanceleaved Stonecrop
Outwash plain of the valley. June–August.

*Sedum rhodanthum* Gray
Redpod Stonecrop
Streams and meadows, subalpine and alpine. August–Sept.

## Cuscutaceae Dodder Family

† *Cuscuta approximata* Bab.
Alfalfa Dodder
Parasitic on a variety of hosts such as *Medicago, Lithosperma* and *Grindelia*; native of Europe. June–August.

## Dipsacaceae Teasel Family

† *Knautia arvensis* (L.) Coult.
Field Scabious
Moist disturbed areas south of Jackson; native of Europe. July–August.

## Droseraceae Sundew Family

* *Drosera anglica* Huds.
English Sundew
Rare, shallow water or sedge meadow, south shore of Leigh Lake. July–August.

## Elaeagnaceae Oleaster Family

*Elaeagnus commutata* Bernh. ex Rydb.
Silverberry
Shrub along the Snake River especially at Moose. June–July.

*Shepherdia canadensis* (L.) Nutt.
Russet Buffaloberry
Along the Snake River and around the shores of the valley lakes. June–July.

## Ericaceae Heath Family

*Arctostaphylos uva-ursi* (L.) Spreng.
Bearberry; Kinnikinnick
Along the Snake River and around the shores of some valley lakes; also Avalanche Canyon; circumboreal. July–August.

*Gaultheria humifusa* (Grah.) Rydb.
Alpine Wintergreen
Moist soil on the shores of Bradley, String and Amphitheater Lakes. July–August.

*Kalmia microphylla* (Hook.) Heller
Alpine Laurel
Shores of the valley lakes and some subalpine lakes. July–August.

*Menziesia ferruginea* Smith var. *glabella* (Gray) Peck
False Huckleberry
Moist woods and mountain stream banks up to 8,500 ft. June–July.

*Phyllodoce empetriformis* (Sw.) D. Don
Pink Mountain Heath
Shorelines of alpine lakes above 9,000 ft. July–Sept.

*Phyllodoce glanduliflora* (Hook.) Cov.
Yellow Mountain Heath
Shorelines of alpine lakes above 9,000 ft. July–Sept.

* *Phyllodoce* X *intermedia* (Hook.) Camp
Hybrid Mountain Heath
Hybrid between *P. empetriformis* and *P. glanduliflora.* Occasionally seen at Holly and Amphitheater Lakes. July–Sept.

*Vaccinium cespitosum* Michx.
Dwarf Blueberry
Meadows and slopes; near Colter Bay. June–July.

*Vaccinium globulare* Rydb.
Blue Huckleberry
Coniferous woods, Taggart Creek. June–July.

*Vaccinium membranaceum* Dougl. ex Torrey
Highbush Huckleberry; Big Blueberry
Lower portions of the canyons and around the shores of the valley lakes. June–July.

*Vaccinium occidentale* Gray
Western Bog Blueberry
Bog-like areas in the valley, outlet of Bradley Lake and small pond east of Colter Bay. June–July.

*Vaccinium scoparium* Leib. ex Cov.
Grouse Whortleberry
Valley lakes, canyons and subalpine, shade of coniferous woods. June–July.

## Fabaceae Pea Family

*Astragalus aboriginorum* Richardson [Barneby 1990 uses *A. australis* (L.) Lam.]
Indian Milkvetch
Rocky slopes and meadows; circumboreal. June–August.

*Astragalus agrestis* Dougl. ex G. Don
Field Milkvetch
Moist meadows and stream banks along the Snake River. June–August.

*Astragalus alpinus* L.
Alpine Milkvetch
Subalpine to alpine; woods, slopes and meadows; circumboreal. July–August.

*Astragalus argophyllus* Nutt.
Silverleaved Milkvetch
Meadows, woods and stream banks in the valley. June–August.

*Astragalus bisulcatus* (Hook.) Gray var. *bisulcatus*
Two-grooved Milkvetch
Hills along the east boundary of the park; poisonous to grazing animals because of the selenium it absorbs from the soil. June–August.

*Astragalus canadensis* L. var. *canadensis*
Canada Milkvetch
Stream banks and meadows along the Snake River; poisonous to livestock. July–August.

*Astragalus cibarius* Sheld.
Browse Milkvetch
Dry morainal soil, northwest of Moose; potentially poisonous to livestock because of nitro-toxins. June–July.

*Astragalus diversifolius* Gray
Meadow Milkvetch
A. Nelson (1086) collection of 1894, Gros Ventre River.

*Astragalus eucosmus* Robins.
Stream banks, meadows and woods, Spread Creek, east side of valley. June–July.

*Astragalus kentrophyta* Gray var. *tegetarius* (Wats.) Dorn
Kentrophyta
Exposed rocky ridges in subalpine and alpine. July–Sept.

*Astragalus miser* Dougl. var. *decumbens* (Nutt. ex T. & G.) Cronq.;
var. *hylophilus* (Rydb.) Barneby; var. *tenuifolius* (Nutt.) Barneby
Weedy Milkvetch
Woods and slopes along the Snake River; poisonous to livestock. June–August.

*Astragalus purshii* Dougl. ex Hook.
Pursh Milkvetch
Meadows along Oxbow Bend of the Snake River. June–July.

* *Astragalus shultziorum* Barneby
Shultz Milkvetch
Subalpine and alpine, Rendezvous Mt. south from summit of aerial tram; Dorn and Lichvar maintain that this species is conspecific with *A. molybdenus*; Marriott lists this species as a narrow endemic. June–July.

*Astragalus tenellus* Pursh
Pulse Milkvetch
Merrill & Wilcox (964) collection of 1901, near Adams Ranch in Jackson Hole. June–July.

* *Astragalus terminalis* Wats.
Railhead Milkvetch
Dry terraces along the Snake River. June–August.

† *Caragana arborescens* Lam.
Siberian Pea-shrub
Remaining from cultivation east of Jackson Lake Dam; native of Siberia. June.

*Glycyrrhiza lepidota* Pursh
Wild Licorice
Along the Snake River bottoms especially below Moose. June–August.

*Hedysarum alpinum* L. var. *americanum* Michx.
Alpine Sweetvetch
Woods, meadows and stream banks. June–July.

*Hedysarum boreale* Nutt. var. *boreale*
Northern Sweetvetch
Gravelly soil along Spread Creek. June–July.

*Hedysarum occidentale* Greene
Western Sweetvetch
Upper South Cascade Canyon and slopes of Symmetry Spire. July–August.

† *Lotus tenuis* Wald. & Kit. [*L. corniculatus* L.]
Birdfoot Trefoil
Disturbed sites in the valley; native of Eurasia. June–August.

*Lupinus argenteus* Pursh var. *argenteus*; var. *depressus* (Rydb.) Hitchc.
Silver Lupine
Shade of coniferous forest throughout the valley; poisonous to livestock especially in the seed stage. June–August.

*Lupinus lepidus* Dougl. ex Lindl. ssp. *caespitosus* (Nutt.) Detling
Dwarf Lupine
Woods, meadows and stream banks in the valley. June–July.

*Lupinus polyphyllus* Lindl.
Meadow Lupine
Woods and stream banks in the valley. June–July.

*Lupinus sericeus* Pursh
Silky Lupine
Dry to mesic sites, often open woods in the valley; poisonous to livestock especially in the seed stage. June–August.

*Lupinus wyethii* Wats.
Wyeth Lupine
Plains, hills and slopes, reported from Togwotee Pass. June–August.

† *Medicago lupulina* L.
Black Medic
Weed of disturbed sites; native of Europe. June–Sept.

† *Medicago sativa* L.
Common Alfalfa
An escape from cultivation throughout the valley; native of Eurasia. June–Sept.

† *Melilotus alba* Medic.
Whitesweet Clover
Weed of disturbed sites; native of Eurasia. June–Sept.

† *Melilotus officinalis* (L.) Pallas
Yellow Sweetclover
Weed of disturbed sites; native of Europe. June–Sept.

† *Onobrychis viciifolia* Scop.
Common Sainfoin
Introduced forage plant, escaping and occupying disturbance sites; native of Europe; first seen at Teton Science School in 1974. June–July.

*Oxytropis campestris* (L.) DC.
Field Locoweed
Limestone, head of Death Canyon, alpine; circumboreal. June–August.

*Oxytropis deflexa* (Pallas) DC.
var. *foliolosa* (Hook.) Barneby: mostly alpine. July–August.
var. *sericea* T. & G.: along the Snake River. July.
Nodding Locoweed

*Oxytropis lagopus* Nutt.
Hills and rocky slopes in the valley. June–August.

*Oxytropis sericea* Nutt. var. *spicata* (Hook.) Barneby
White Locoweed
Meadows and slopes, near mouth of Fish Creek, Gros Ventre drainage. June–July.

*Oxytropis viscida* Nutt.
Sticky Oxytrope
Hills and slopes along the Gros Ventre River and Sportsman Ridge (T40N, R112W, S10). July–August.

† *Trifolium hybridum* L.
Alsike Clover
Weed of disturbed sites; native of Eurasia. June–Sept.

*Trifolium longipes* Nutt. var. *reflexum* A. Nels.
Summer Clover
Meadows and stream banks east of Jackson Lake. June–August.

† *Trifolium pratense* L.
Red Clover
Cultivated forage plant, escaping to disturbed sites; native of Eurasia. June–Sept.

† *Trifolium repens* L.
White Clover
Cultivated forage plant, escaping to disturbed sites; native of Eurasia. June–August.

*Vicia americana* Muhl. ex Willd. var. *americana*
American Vetch
Woods and meadows along the Snake River; an extremely variable species. June–August.

## Fumariaceae Fumitory Family

*Corydalis aurea* Willd.
Golden Corydalis
Moist to dry soil along the Snake River. June–August.

* *Dicentra uniflora* Kell.
Steershead
Open dry to moist sites in the valley, appearing as the snow melts. April–June.

## Gentianaceae Gentian Family

*Frasera speciosa* Dougl. ex Griseb.
Green Gentian; Elkweed
From the valley to the alpine; open slopes or open woods. July–August.

*Gentiana affinis* Griseb. var. *affinis*
Rocky Mountain Gentian
Mesic sites along the Snake River. July–August.

*Gentiana calycosa* Griseb.
Explorers Gentian
Meadows and stream banks in the subalpine. July–August.

*Gentiana prostrata* Haenke
Moss Gentian
East of Gros Ventre Slide (Reed 2701 in 1949). June–August.

*Gentianella amarella* (L.) Boerner
Felwort
Moist areas and stream banks, north side of Signal Mt., also south of Colter Bay Village. July–August.

*Gentianella barbellata* (Engelm.) Gillett
Stiffhair Gentian
Woods and meadows in the mountains, (Crystal Creek Divide, Murie 1052, 1933). August.

*Gentianella detonsa* (Rottb.) G. Don var. *elegans* (A. Nels.) Dorn
Smaller Fringed Gentian
Moist areas, willow patches north of Jackson Lake Lodge. July–Sept.

*Swertia perennis* L.
Swertia; Felwort
Wet meadows and stream banks in canyons such as Death and Webb canyons; circumboreal. July–August.

## Geraniaceae Geranium Family

† *Erodium cicutarium* (L.) L'Her. ex Aiton
Storksbill; Filaree
Open disturbed sites; native of Europe. April–Sept.

*Geranium richardsonii* Fisch. & Trautv.
Richardson Geranium
Stream banks or moist meadows throughout the valley. June–Sept.

*Geranium viscosissimum* F. & M.
Sticky Geranium
Along the Snake River and open xeric sites of the valley moraines. June–August.

## Grossulariaceae Currant Family

*Ribes aureum* Pursh
Golden Currant
Moist sites along the Snake River. May–June.

*Ribes cereum* Dougl.
Squaw Currant
Open woods and rocky areas, north end of Blacktail Butte. May–July.

*Ribes hudsonianum* Richardson
Western Black Currant
Moist sites in the major canyons. June–July.

*Ribes inerme* Rydb.
Whitestem Gooseberry
Moist sites and stream banks along the Snake River. June–July.

*Ribes lacustre* (Pers.) Poiret
Prickly Currant
Moist woods and stream banks of the canyons and valley. June–July.

*Ribes montigenum* McClatchie
Gooseberry Currant
Subalpine or alpine, rocky sites. July–August.

*Ribes oxyacanthoides* L. var. *setosum* (Lindl.) Dorn
Missouri Gooseberry
Stream banks of the valley. June–July.

† *Ribes rubrum* L.
Red Currant
Disturbed areas. Cultivated; native of Europe. June.

*Ribes viscosissimum* Pursh
Sticky Currant
Moist areas, north slope of Signal Mt. and the shore of Bradley Lake. June–July.

## Haloragaceae Water Milfoil Family

*Myriophyllum sibiricum* Kom. [*M. exalbescens*]
Water Milfoil
Quiet streams, ponds and lakes; circumboreal. June–August.

## Hippuridaceae Marestail Family

*Hippuris vulgaris* L.
Marestail
Submerged in water, shallow ponds north of Jackson Lake Dam; circumboreal. June–August.

## Hydrophyllaceae Waterleaf Family

*Hydrophyllum capitatum* Dougl. ex Benth. var. *capitatum*
Ballhead Waterleaf
Woods and slopes of the valley moraines. May–July.

*Nemophila breviflora* Gray
Great Basin Nemophila
Along the Snake River near Menor's Ferry. May–June.

*Phacelia franklinii* (R. Br.) Gray
Franklin Phacelia
Moist soil along the Snake River. June–August.

*Phacelia hastata* Dougl. ex Lehm.
Cordilleran Phacelia
Talus slopes of the subalpine. July–Sept.

*Phacelia heterophylla* Pursh var. *virgata* (Greene) Dorn
Wand Phacelia
Open or disturbed sites in the valley. June–August.

*Phacelia linearis* (Pursh) Holz.
Narrow-leaved Phacelia
Woods and slopes along the Snake River. May–July.

*Phacelia sericea* (Grah. ex Hook.) Gray
Silky Phacelia
Rocky sites in the canyons. July–August.

## Lamiaceae Mint Family

*Agastache urticifolia* (Benth.) Kuntze
Giant Hyssop
Open slopes of valley moraines. June–August.

*Dracocephalum parviflorum* Nutt.
Smallflowered Dragonhead
Open moist sites in the valley, often weedy. July–August.

*Mentha arvensis* L. var. *canadensis* (L.) Kuntze
Field Mint
Moist areas, Snake River and small ponds north of Colter Bay; circumboreal. June–Sept.

† *Nepeta cataria* L.
Catnip
Weed of disturbed sites, Kelly Warm Spring; native of Eurasia. June–Sept.

*Physostegia parviflora* Nutt. ex Gray
False Dragonhead
Wet place along the Snake River (one collection C. Mernin 1979). June–August.

*Prunella vulgaris* L.
Common Selfheal
Moist areas beside streams and lakes of the valley; nearly cosmopolitan. June–August.

*Scutellaria galericulata* L.
Skullcap
Wet meadows north of Jackson Lake Dam; circumboreal. June–August.

*Stachys palustris* L. var. *pilosa* (Nutt.) Fern.
Hedge Nettle
Moist sites, reported 20 miles south of Jackson (Merrill & Wilcox 953 in 1901); circumboreal. June–August.

## Lentibulariaceae Bladderwort Family

*Utricularia minor* L.
Lesser Bladderwort
Rarely seen, standing or slowly moving water, south side of Leigh Lake. July–August.

*Utricularia vulgaris* L.
Common Bladderwort
Submerged in water, ponds north of Jackson Lake Lodge; circumboreal. July–August.

## Limnanthaceae Meadowfoam Family

*Floerkea proserpinacoides* Willd.
False Mermaid
Moist soil, especially under shrubs in the valley. June–July.

## Linaceae Flax Family

*Linum lewisii* Pursh
Wild Flax
Outwash plain of the valley to alpine zone. June–August.

## Loasaceae Loasa Family

*Mentzelia dispersa* Wats.
Bush Blazing Star
Dry rocky slopes throughout the valley. June–August.

*Mentzelia laevicaulis* (Dougl. ex Hook.) T. & G.
Smoothstem Blazing Star
Gravelly soil, particulary roadside cuts. June–Sept.

## Malvaceae Mallow Family

*Iliamna rivularis* (Dougl. ex Hook.) Greene
Wild Hollyhock
Streams, meadows and roadsides; increasing after a fire. July–Sept.

† *Malva neglecta* Wallr.
Roundleaf Mallow
Weed of disturbed sites; native of Europe. June–Sept.

*Sidalcea oregana* (Nutt. ex T. & G.) Gray
Oregon Checkermallow
Meadows and open woods, vicinity of Hoback Canyon (Williams 836, 1932). July–August.

## Menyanthaceae Buck Bean Family

*Menyanthes trifoliata* L.
Buck Bean
Shallow ponds and lakes, northeast portions of the park; circumboreal. July–August.

## Monotropaceae Indian Pipe Family

*Hypopitys monotropa* Crantz
Pinesap
Shade of coniferous forest; west side of Leigh Lake and east shore of String Lake; mycotrophic. July–August.

*Pterospora andromedea* Nutt.
Pinedrops
Shade of coniferous forests of the valley; mycotrophic. June–August.

## Nymphaceae Waterlily Family

*Nuphar polysepalum* Engelm.
Rocky Mt. Yellow Waterlily
Aquatic in ponds and shallow lakes of the valley. July–August.

## Oleaceae Olive Family

† *Syringa vulgaris* L.
Common Lilac
Cultivated shrub on some ranches in the valley; native of Europe. June–July.

## Onagraceae Evening Primrose Family

*Camissonia breviflora* (T. & G.) Raven
Few-flowered Camissonia
Stream banks and drying ponds in the north end of the valley. June–August.

*Camissonia subacaulis* (Pursh) Raven
Stemless Camissonia
Meadows in the valley, usually where dry by late summer. July–August.

*Circaea alpina* L.
Enchanters Nightshade
Moist woods, Death Canyon Trail at Phelps Lake level. June–August.

*Clarkia pulchella* Pursh
Clarkia
Rocky slopes along Snake River; not seen since 1892 (F. McCoullough, specimen at University of Wyoming.)

*Epilobium alpinum* L. Includes the following intergrading taxa:
[*E. anagallidifolium* Lam.; *E. clavatum* Trel.; *E. lactiflorum* Hausskn.; *E. hornemannii* Reichb.]
Alpine Willow-herb
Moist sites in the subalpine and alpine. June–August.

*Epilobium angustifolium* L.
Fireweed
Valley to subalpine; various habitats; invading after forest fires; circumboreal. July–Sept.

*Epilobium brachycarpum* Presl [*E. paniculatum* Nutt.]
Panicled Willow-herb
Sagebrush and open woods, Blacktail Butte and slopes of Signal Mt.; a complex and highly variable species. July–August.

*Epilobium canum* (Greene) Raven [*Zauschneria garrettii* Nels.]
Firechalice
Rocky ledges in such canyons as Death, Waterfalls and Cascade, up to 8,000 ft. July–Sept.

*Epilobium ciliatum* Raf. var. *glandulosum* (Lehm.) Dorn
American Willow-herb
Moist woods along the Snake River. June–August.

*Epilobium latifolium* L.
Broad-leaved Willow-herb
Stream banks in Garnet and South Cascade Canyons. July–August.

*Epilobium suffruticosum* Nutt.
Shrubby Willow-herb
Gravel of Snake River bottoms, especially 1/2 mile below Jackson Lake Dam, south side of river. July–August.

*Gayophytum diffusum* T. & G. var. *diffusum*
Diffuse Groundsmoke
Dry gravelly soils in the valley. June–August.

*Oenothera cespitosa* Nutt. var. *cespitosa*
Tufted Evening Primrose
Roadside cuts and dry hills of the valley. June–July.

*Oenothera flava* (A. Nels.) Garrett
Long-tubed Evening Primrose
Moist, poorly drained soil along the Snake River. July–August.

*Oenothera pallida* Lindl.
Pale Evening Primrose
Sandy or gravelly soils, west side of Blacktail Butte. June–July.

*Oenothera villosa* Thunb. var. *strigosa* (Rydb.) Dorn [*O. biennis* L.]
Common Evening Primrose
Disturbed sites in the valley. June–August.

## Orobanchaceae Broomrape Family

*Orobanche corymbosa* (Rydb.) Ferris
Flat-topped Broomrape
Valley; often parasitic on *Artemisia* spp. June–August.

*Orobanche fasciculata* Nutt.
Clustered Broomrape
Sagebrush vegetation; parasitic on several spp., especially *Artemisia* spp. June–July.

* *Orobanche ludoviciana* Nutt. var. *arenosa* (Suksd.) Cronq.
Louisiana Broomrape
Sagebrush vegetation; often parasitic on *Artemisia* spp. June–July.

*Orobanche uniflora* L.
Naked Broomrape
Many habitats; often parasitic on a variety of hosts. June–August.

## Oxalidaceae Woodsorrel Family

*Oxalis dillenii* Jacq.
Dillens Woodsorrel
Disturbed sites along the Snake River. June–August.

## Paeoniaceae Peony Family

* *Paeonia brownii* Dougl. ex Hook.
Browns Peony
Rare at the south end of Jenny Lake, near boat dock, also at junction of Taggart Creek and Teton Park Rd. June.

## Plantaginaceae Plantain Family

† *Plantago lanceolata* L.
English Plantain
Weed of disturbed sites; native of Eurasia. June–August.

† *Plantago major* L.
Common Plantain
Weed of disturbed sites; native of Europe. June–August.

*Plantago tweedyi* Gray
Tweedy Plantain
Moist places in the mountains, vicinity of Teton Pass. June–Sept.

## Polemoniaceae Phlox Family

*Collomia linearis* Nutt.
Narrowleaf Collomia
Sagebrush flats and moraines throughout the valley. May–July.

*Gilia tenerrima* Gray
Soft Gilia
Open woods, near Moose and Teton Canyon (11 miles east of Driggs, Idaho). June–August.

*Gymnosteris parvula* Heller
Small Gymnosteris
Dry to mesic sites between Jenny Lake and the Snake River. May–July.

*Ipomopsis aggregata* (Pursh) Grant [*Gilia aggregata*]
Scarlet Gilia
Valley to subalpine, open rocky slopes and dry meadows. June–August.

*Ipomopsis spicata* (Nutt.) Grant var. *spicata*
Spike Gilia
Reported by R. Lichvar (263 and 977); Sportsmans Ridge and Curtis Canyon Campground. June–July.

*Leptodactylon pungens* (Torrey) Nutt.
Common Prickly Phlox
Sagebrush flats south of Signal Mt. June–July.

*Linanthastrum nuttallii* (Gray) Ewan [*Linanthus nuttallii* (Gray) Greene]
Nuttall Gilia
Meadows and slopes in the valley. June–August.

*Linanthus septentrionalis* Mason
Northern Linanthus
Dry meadows and open woods along the Snake River. June–July.

*Microsteris gracilis* (Hook.) Greene var. *gracilis*
Microsteris
Hills and slopes throughout the valley. May–June.

*Navarretia breweri* (Gray) Greene
Yellow Navarretia
Gros Ventre River; one collection, August 1894, A. Nelson (1094). August.

*Phlox hoodii* Richardson
Carpet Phlox
Sagebrush flats of the valley. May–June.

*Phlox longifolia* Nutt.
Longleaf Phlox
Sagebrush flats surrounding the airport. May–July.

*Phlox multiflora* A. Nels.
Rocky Mountain Phlox
Valley to subalpine, coniferous woods to open rocky ridges. June–August.

*Phlox pulvinata* (Wherry) Cronq.
Cushion Phlox
Subalpine to alpine, open rocky sites. June–August.

*Polemonium occidentale* Greene
Western Jacobs Ladder
Wet meadows, east of Jackson Lake Lodge and National Elk Refuge. June–August.

*Polemonium pulcherrimum* Hook.
Short Jacobs Ladder
Midmontane and subalpine, moist or shaded sites. June–August.

*Polemonium viscosum* Nutt.
Sky Pilot
Subalpine and alpine, open rocky sites. July–August.

## Polygonaceae Buckwheat Family

*Eriogonum brevicaule* Nutt.
Shortstem Buckwheat
Wind swept ridges to talus, Gros Ventre Slide. June–August.

*Eriogonum caespitosum* Nutt.
Mat Buckwheat
Hills and rocky sites, 1.5 miles east of Elk Ranch Reservoir. June–August.

*Eriogonum cernuum* Nutt.
Nodding Buckwheat
Dry sites, pothole area south of Signal Mt. June–August.

*Eriogonum heracleoides* Nutt.
Whorled Buckwheat
Open rocky slopes in the valley. June–July.

*Eriogonum ovalifolium* Nutt. var. *ochroleucum* (Small ex Rydb.) Peck
Cushion Buckwheat
Rocky slopes and ridges, Garnet Canyon. July–August.

*Eriogonum umbellatum* Torrey
Sulfur Buckwheat
Sagebrush to coniferous forest in the valley; a highly variable species. June–August.

*Oxyria digyna* (L.) Hill
Mountain Sorrel
Subalpine and alpine, talus slopes; circumboreal. July–Sept.

† *Polygonum achoreum* Blake
Knotweed
Disturbed sites, Oxbow Bend of Snake River; origin unknown. June–August.

*Polygonum amphibium* L. var. *stipulaceum* Coleman
Water Smartweed
Submerged in water or on mud flats, north end of the valley. June–August.

*Polygonum aviculare* L. [*P. arenastrum*]
Knotweed
Disturbed sites; circumboreal. June–August.

*Polygonum bistortoides* Pursh
American Bistort
Wet meadows and stream banks from the valley to the alpine. June–August.

† *Polygonum convolvulus* L.
Wild Buckwheat
Weed of disturbed sites; native of Europe. July–August.

*Polygonum douglasii* Greene var. *austiniae* (Greene) Jones
Douglas Knotweed
Rocky sites above Graul's Mine, 3 miles up Webb Canyon, also in disturbed sites of the valley. July–Sept.

*Polygonum kelloggii* Greene var. *confertiflorum* (Nutt. ex Piper) Dorn
Kellogg Knotweed
Seasonally moist areas in the valley and the mountains. June–Sept.

*Polygonum minimum* Wats.
Broadleaf Knotweed
Rocky sites or talus in the mountains, Symmetry Spire, also Webb Canyon. July–August.

*Polygonum sawatchense* Small
Sawatch Knotweed
Moist places, Togwotee Pass. July–August.

*Polygonum viviparum* L.
Alpine Bistort
Subalpine and alpine; moist meadows; lower flowers replaced by bulblets; circumboreal. July–Sept.

*Polygonum watsonii* Small
Watson Knotweed
Moist areas, shores of Jackson Lake and 2 miles south of Moran. July–August.

† *Rheum rhubarbarum* L.
Rhubarb
Remaining from cultivation, 1/4 to 1/2 mile north of the Highlands; native of Eurasia. Leaf blades contain poisonous oxalic acid or soluble oxalates. June.

† *Rumex acetosella* L.
Sheep Sorrel
Weed of disturbed sites; native of Eurasia. July–Sept.

† *Rumex crispus* L.
Curly Dock
Weed of disturbed sites; native of Eurasia. June–August.

*Rumex maritimus* L. var. *fueginus* (Phil.) Dusen
Golden Dock
Moist sites along the Snake River. July–Sept.

*Rumex paucifolius* Nutt.
Alpine Sorrel
Midmontane and subalpine; meadows. July–August.

*Rumex salicifolius* Weinm. ssp. *triangulivalvis* Danser
Willowleaf Dock
Moist or dry sites, especially along roadsides. June–August.

*Rumex venosus* Pursh
Veiny Dock
Not seen in Grand Teton National Park, but reported in adjacent areas. Large fruits and fleshy leaves. June.

## Portulacaceae Purslane Family

*Claytonia lanceolata* Pursh var. *lanceolata*
Springbeauty
Sagebrush flats in the valley to alpine slopes. May–August.

*Claytonia megarhiza* (Gray) Parry ex Wats.
Alpine Springbeauty
Rare on talus slopes, Mt. Moran and Buck Mt., above 10,000 ft. July–August.

*Lewisia pygmaea* (Gray) Robins.
Least Lewisia
Sagebrush north of Timbered Island to meadows in the alpine. June–August.

*Lewisia triphylla* (Wats.) Robins.
Threeleaf Lewisia
Moist sites, 1/4 mile below Hidden Falls, also in Alaska Basin. June–August.

*Montia chamissoi* (Ledeb. ex Spreng.) Greene
Water Springbeauty
Wet meadows and exposed wet slopes. June–July.

*Spraguea umbellata* Torr. var. *caudicifera* Gray
Pussy Paws
Rocky places along the Skyline Trail south of Static Peak. July–August.

## Primulaceae Primrose Family

*Androsace filiformis* Retz.
Slender Rock-jasmine
From the valley to the alpine, moist sites. June–August.

*Androsace septentrionalis* L. var. *subulifera*
Pygmyflower
Moist or dry areas, montane to alpine. June–August.

*Dodecatheon conjugens* Greene
Slimpod Shooting Star
Sagebrush flats, especially in vicinity of Jenny Lake. June.

*Dodecatheon jeffreyi* van Houtte
Mountain Shooting Star
Moist sites; reported in Waterfalls Canyon. July–August.

*Dodecatheon pulchellum* (Raf.) Merr.
Beautiful Shooting Star
Valley to subalpine; meadows and streamsides. June–July.

*Primula incana* Jones
American Primrose
Stream banks and meadows; Soda Lake (S11, T41N, R112W) Lichvar (937). July–August.

*Primula parryi* Gray
Parry Primrose
Alpine to subalpine; moist sites. July–Sept.

## Pyrolaceae Wintergreen Family

*Chimaphila umbellata* (L.) Bart. var. *occidentalis* (Rydb.) Blake
Common Pipsissewa
Shade of coniferous forests on the shores of the valley lakes. June–August.

*Moneses uniflora* (L.) Gray
Woodnymph
Rare around the shores of the valley lakes and extending into some canyons. July–August.

*Pyrola asarifolia* Michx.
Liver-leaf Wintergreen
Moist sites in coniferous forests of the valley. June–August.

*Pyrola chlorantha* Sw.
Green Shinleaf
Moist sites in woods, Bradley Creek and north slope of Signal Mt. July–August.

*Pyrola minor* L.
Lesser Wintergreen
Midmontane; coniferous woods; moist sites. June–August.

* *Pyrola picta* Smith var. *dentata* (Smith) Dorn
White Vein Pyrola
Midmontane; coniferous woods. July–August.

*Pyrola secunda* L. [*Orthilia s.*]
One-sided Wintergreen
Coniferous woods around the shores of the valley lakes. June–August.

## Ranunculaceae Buttercup Family

*Aconitum columbianum* Nutt.
Columbia Monkshood
Moist woods and stream banks, in the valley and extending into the canyons. Because of several alkaloids, all parts of this plant are poisonous to humans. June–Sept.

*Actaea rubra* (Aiton) Willd.
Western Baneberry
Moist sites in the woods of the valley and canyons. Poisonous to humans; toxic action is attributed to a glycoside. Only 6 berries can cause severe symptoms. Both red and white berries occur in this species. June–July.

*Anemone multifida* Poiret
Globeflower
Open woods and meadows along the Snake River. July–August.

* *Anemone parviflora* Michx.
Small-flowered Anemone
Moist areas in the mountains; south end of Teton Range above 9,500 ft. July–August.

*Anemone patens* L. var. *multifida* Pritzel
Pasque Flower
Open woods, Snow King Mt. above Jackson. May–June.

*Anemone tetonensis* Porter ex Britt.
Teton Anemone
Alpine and subalpine, rocky sites. July–August.

*Aquilegia coerulea* James
Colorado Columbine
Meadows and stream banks in the subalpine. July–August.

*Aquilegia flavescens* Wats.
Yellow Columbine
Moist mountain meadows in the subalpine. June–August.

*Caltha leptosepala* DC.
Marsh Marigold
Wet meadows in the subalpine and alpine. June–August.

*Clematis hirsutissima* Pursh
Sugarbowl
Sagebrush to spruce-fir, valley and canyons. June–July.

*Clematis ligusticifolia* Nutt.
Western Virgins-bower
Lower portion of Death Canyon below 7,800 ft.; climbing vine on trees. June–July.

*Clematis occidentalis* (Hornem.) DC. var. *grosseserrata* (Rydb.) Pringle
Western Clematis
Open woods in Cascade Canyon and along Pilgrim Creek. June–July.

*Delphinium burkei* Greene
Meadow Larkspur
Wet meadows, north drainage of Pilgrim Creek; Beetle (13974). July–August.

*Delphinium nuttallianum* Pritzel ex Walpers
Nuttall Larkspur
Gravelly soils throughout the valley. Larkspurs spp. contain several alkaloids that are poisonous to livestock and humans, especially young plants and mature seeds. May–June.

*Delphinium occidentale* Wats.
Duncecap Larkspur
Meadows, stream banks and open woods in the valley. Poisonous. June–August.

*Myosurus minimus L.*
Tiny Mousetail
Moist sites, especially in vernal pools along the Snake River; circumboreal. May–June.

*Ranunculus acriformis* Gray
Sharp Buttercup
Stream banks and meadows north and east of Jackson Lake Dam. July–August.

*Ranunculus aquatilis* L. var. *capillaceus* DC.
Water Buttercup
Submerged in pools and sluggish streams throughout the valley. July–August.

*Ranunculus cymbalaria* Pursh var. *alpinus* Hook.
Marsh Buttercup
Wet places along the Snake River. June–August.

*Ranunculus eschscholtzii* Schlecht.
Often separated into three vars. based on intergrading leaf differences.
Eschscholtz Buttercup
Subalpine and alpine; meadows near melting snowbanks. June–August.

*Ranunculus flabellaris* Raf.
Yellow Water Buttercup
Pond north of Berry Creek Cabin. June–July.

*Ranunculus flammula* L. var. *filiformis* (Michx.) Hook.
Creeping Spearwort
Margins of shallow ponds north of Colter Bay Village; circumboreal. June–August.

*Ranunculus glaberrimus* Hook. var. *ellipticus* Greene
Sagebrush Buttercup
Sagebrush and open woods of the valley. June–July.

*Ranunculus gmelinii* DC. var. *purshii* (Richardson) Hara
Gmelin Buttercup
Ponds and mud flats, Lower Slide Lake. July–August.

*Ranunculus inamoenus* Greene
Pleasant Buttercup
Sagebrush and open woods along the Snake River. June–July.

*Ranunculus jovis* A. Nels.
Utah Buttercup
Sagebrush flats of the valley. April–June.

*Ranunculus macounii* Britt.
Macoun Buttercup
Irrigated meadows on the Elk Ranch. June–July.

*Ranunculus natans* Meyer var. *intertextus* (Greene) Benson
Floating Buttercup
Ponds and stream banks along the Snake River. July–August.

* *Ranunculus orthorhynchus* Hook. var. *platyphyllus* Gray
Straightbeak Buttercup
Moist meadows and stream banks in the valley. June–August.

† *Ranunculus repens* L. var. *pleniflorus* Fern.
Creeping Buttercup
Moist disturbed sites; native of Europe. June–August.

*Ranunculus sceleratus* L. var. *multifidus* Nutt.
Blister Buttercup
Moist areas or in water along the Snake River. June–August.

*Ranunculus uncinatus* D. Don
Little Buttercup
Moist areas along the Snake River. June–August.

*Thalictrum fendleri* Engelm. ex Gray
Fendler Meadowrue
Stream banks in the coniferous forests of the valley. June–July.

*Thalictrum occidentale* Gray
Western Meadowrue
Moist sites along the Snake River. June–July.

*Thalictrum sparsiflorum* Turcz. ex Fisch. & Meyer var. *saximontanum* Boivin
Few-flowered Meadowrue
Damp woods, west shore of Jenny Lake. June–July.

*Trollius laxus* Salisb. var. *albiflorus* Gray
American Globeflower
Wet meadows in the subalpine and alpine. July–August.

## Rhamnaceae Buckthorn Family

*Ceanothus velutinus* Dougl. ex Hook.
Snowbrush Ceanothus
Woods, around the shores of the valley lakes. June–August.

*Rhamnus alnifolia* L'Her.
Alder Buckthorn
Moist sites along the Snake River and the east shore of Jackson Lake. June–August.

## Rosaceae Rose Family

*Amelanchier alnifolia* (Nutt.) Nutt. ex Roem.
Serviceberry
Along the Snake River and around the shores of the valley lakes. Fruits are edible. June.

*Cercocarpus ledifolius* Nutt. var. *ledifolius*; var. *intercedens* Schneider
Curl-leaf Mountain Mahogany
Mountain slopes, Granite and Darby Canyons. June.

*Crataegus douglasii* Lindl. var. *douglasii*; var. *rivularis* (Nutt.) Sarg.
Douglas Hawthorn
Moist sites along the Snake River and Taggart Lake Trail. June.

*Dryas octopetala* L. var. *hookeriana* (Juz.) Breit.
Dryad; Mountain Avens
Talus and exposed ridges in the alpine, Skyline Trail. July–August.

*Fragaria vesca* L.
Woodland Strawberry
Woods, stream banks and meadows in the valley. June–July.

*Fragaria virginiana* Miller
Virginia Strawberry
Woods and meadows along the Snake River, Oxbow Bend. June–July.

*Geum macrophyllum* Willd. var. *perincisum* (Rydb.) Raup
Largeleaf Avens
Stream banks, meadows and moist woods of the valley. June–August.

*Geum rossii* (R. Br.) Ser. var. *turbinatum* (Rydb.) Hitchc.
Alpine Avens
Alpine and subalpine, moist rocky sites. July–August.

*Geum triflorum* Pursh
Prairie Smoke
Sagebrush flats and meadows of the valley. June–July.

*Ivesia gordonii* (Hook.) T. & G.
Gordon Ivesia
Occasionally along stream banks in the valley, but usually in rocky areas of the alpine. June–August.

*Pentaphylloides floribunda* (Pursh) Löve [*Potentilla fruticosa* L.]
Shrubby Cinquefoil
Moist areas along the Snake River and into the canyons. June–August.

*Petrophytum caespitosum* (Nutt.) Rydb.
Rock Spiraea
Limestone outcroppings in the valley, north end of Blacktail Butte and 2 miles up Berry Creek. August–Sept.

*Physocarpus malvaceus* (Greene) Kuntze
Mallow Ninebark
Introduced in plantings at the Jackson Lake Lodge and reported by Dorn as occurring in northwest Wyoming. June–July.

*Potentilla anserina* L.
Silverweed Cinquefoil
Mud flats and drying ponds, north end of Jackson Lake. June–July.

*Potentilla arguta* Pursh
Tall Cinquefoil
Meadows along the Snake River. June–August.

*Potentilla brevifolia* Nutt. ex T. & G.
Small-leaf Cinquefoil
Open slopes in the alpine. July–August.

*Potentilla concinna* Richardson
Elegant Cinquefoil
Rocky sites in alpine zone, Static Peak and Table Mt. July–August.

*Potentilla diversifolia* Lehm.
Varileaf Cinquefoil
Meadows and rocky slopes in the alpine. July–August.

*Potentilla fissa* Nutt.
Splitleaf Cinquefoil
Slopes in Cascade Canyon. July–August.

* *Potentilla flabellifolia* Hook. ex T. & G.
Fan-leaf Cinquefoil
Meadows and stream banks, Holly Lake Cirque. July–August.

*Potentilla glandulosa* Lindl. var. *pseudorupestris* (Rydb.) Breit.
Gland Cinquefoil
Many plant communities in the canyons. June–August.

*Potentilla gracilis* Dougl. ex Hook.
Showy Cinquefoil
A variable species subdivided into several varieties based on some tenuous leaf and trichome characters; meadows and stream banks. June–August.

*Potentilla hookeriana* Lehm.
Hooker Cinquefoil
Rocky places in the mountains, vicinity of Hoback Canyon. July–August.

* *Potentilla nivea* L.
Snow Cinquefoil
South facing slopes in the alpine. July–August.

*Potentilla norvegica* L.
Norwegian Cinquefoil
Moist, disturbed areas along the Snake River. June–August.

*Potentilla ovina* Macoun
Sheep Cinquefoil
Rocky slopes and meadows in the subalpine and alpine. July–August.

*Potentilla palustris* (L.) Scop.
Purple Cinquefoil
Lake shores, Arizona Lake and Bradley Lake. July–August.

*Prunus virginiana* L. var. *melanocarpa* (A. Nels.) Sarg.
Chokecherry
Along the Snake River and on the moraines of the valley. Bark, leaves and seeds are poisonous because of toxins which break down to prussic acid when hydrolized. June.

*Purshia tridentata* (Pursh) DC.
Antelope Bitterbrush
Sagebrush flats, particularly south of Park Headquarters. June.

† *Pyrus malus* L.
Apple
Kelly Warm Spring; native of Eurasia; escaped. June.

*Rosa nutkana* Presl var. *hispida* Fern.
Nutka Rose
Wooded or open slopes in the valley. June–July.

*Rosa woodsii* Lindl.
Woods Rose
Streamsides in the valley. June–July.

*Rubus idaeus* L.
Red Raspberry
Along the Snake River and talus slopes in the canyons. June–July.

*Rubus laciniatus* Willd.
Evergreen Blackberry
Gravelly banks of Snake River below the Jackson Lake Dam; a possible escaped cultigen. June.

*Rubus parviflorus* Nutt.
Western Thimbleberry
Wooded areas in the valley and the canyons. June–July.

*Sibbaldia procumbens* L.
Prostrate Sibbaldia
Moist sites below melting snowbanks, subalpine and alpine; circumboreal. July–August.

*Sorbus scopulina* Greene
Rocky Mountain Ash
Lower slopes of the mountains and along the shores of the valley lakes. June–July.

*Spiraea betulifolia* Pallas var. *lucida* (Dougl.) Hitchc.
Shinyleaf Spiraea
Moraines and shores of the valley lakes. June–July.

*Spiraea splendens* Baumann ex Koch [*S. densiflora* Nutt.]
Subalpine Spiraea
Mountain slopes and in the canyons up to 9,000 ft. June–July.

## Rubiaceae Madder Family

*Galium aparine* L.
Catchweed Bedstraw
Weedy species in a variety of habitats in the valley; circumboreal. June–July.

*Galium bifolium* Wats.
Twinleaf Bedstraw
Under aspen along Snake River and extending into canyons. June–August.

*Galium boreale* L.
Northern Bedstraw
Morainal soils of the valley; circumboreal. June-August.

*Galium trifidum* L.
Small Bedstraw
Moist riparian sites of the valley; circumboreal. June–August.

*Galium triflorum* Michx.
Sweetscented Bedstraw
Moist woods, Teton Canyon and Webb Canyon; circumboreal. July–August.

* *Kelloggia galioides* Torrey
Kelloggia
Rocky sites, east face of Symmetry Spire. July–August.

## Salicaceae Willow Family

*Populus acuminata* Rydb.
Lanceleaf Cottonwood
Along the Gros Ventre River. R. Lichvar (542), 1977. 100 ft. (30m). Used in a broad sense to include hybrids between *P. angustifolia* and broad-leaved taxa. May–June.

*Populus angustifolia* James
Narrowleaf Cottonwood
Common along the Snake River; hybridizes freely with broad-leaved spp. of the genus. 60 ft. (18m). May–June.

*Populus balsamifera* L.
Balsam Poplar
Stream banks and shores in the valley. 100 ft. (30m). May–June.

*Populus tremuloides* Michx.
Quaking Aspen
Moist sites in the valley and canyons. 40–60 ft. (12–18m). May.

*Salix arctica* Pallas
Arctic Willow
Wet meadows and moist rocky slopes in the alpine. July–August.

*Salix bebbiana* Sarg.
Bebb Willow
Stream banks and seepage areas in the valley. May–June.

*Salix boothii* Dorn
Booth Willow
Water courses and wet meadows in the valley. May–June.

*Salix cascadensis* Cock.
Cascade Willow
High peaks and cirques in the alpine. July–August.

*Salix drummondiana* Barratt ex Hook.
Drummond Willow
Water courses and wet meadows in the valley and canyons. May–June.

*Salix eastwoodiae* Cock. ex Heller
Eastwood Willow
Wet meadows and stream banks in the subalpine. June–August.

*Salix exigua* Nutt.
Coyote Willow
Stream banks and ditches in the valley. May–July.

*Salix geyeriana* Anderss.
Geyer Willow
Wet places in the valley. May–June.

*Salix glauca* L. var. *villosa* (Hook.) Anderss.
Glaucous Willow
Stream banks and talus in the subalpine. June–August.

*Salix lasiandra* Benth. var. *caudata* (Nutt.) Sudw.
Whiplash Willow
Stream banks and ponds in the valley. May–June.

*Salix lemmonii* Bebb
Lemmon Willow
Stream banks and wet meadows in the valley. May–June.

*Salix lutea* Nutt.
Yellow Willow
Stream banks and ditches in the valley. May–June.

*Salix melanopsis* Nutt.
Dusky Willow
Stream banks in the valley and canyons. May–June.

*Salix planifolia* Pursh var. *planifolia*; var. *monica* (Bebb) Schneider
Plain-leaf Willow
Stream banks and wet meadows in the mountains. June.

*Salix pseudomonticola* Ball
Mountain Willow
Stream banks in the valley. May–June.

*Salix reticulata* L. var. *nana* Anderss.
Network Willow
Moist sites on the high peaks and cirques in the alpine. July–August.

*Salix rotundifolia* Trautv. var. *dodgeana* (Rydb.) Murray
Roundleaf Willow
Usually on limestone, subalpine and alpine. July–August.

*Salix scouleriana* Barratt ex Hook.
Scouler Willow
Drier woods and clearings in the valley. June.

*Salix tweedyi* (Bebb ex Rose) Ball
Tweedy Willow
Stream banks and wet meadows in the mountains. June–July.

*Salix wolfii* Bebb
Wolf Willow
Stream banks and margins of lakes in the valley. May–June.

## Santalaceae Sandalwood Family

*Comandra umbellata* (L.) Nutt. var. *pallida* (A. DC.) Jones
Bastard Toadflax
Sagebrush communities at such places as Windy Point, Timbered Island and Moose. Plants have extensive rhizomes and are hemiparasitic. Species is also alternate host for a rust of *Pinus contorta—Cronartium comandrae.* June.

## Saxifragaceae Saxifrage Family

*Conimitella williamsii* (Eaton) Rydb.
Conimitella
Rocks and slopes, Hoback Canyon. June.

*Heuchera parvifolia* Nutt. ex T. & G.
Common Alumroot
North facing slopes in rock crevices, trail to Phelps Lake Lookout, also Blacktail Butte. June–August.

*Lithophragma glabrum* Nutt.
Fringecup Woodlandstar
Sagebrush flats in the valley. May–June.

*Lithophragma parviflorum* (Hook.) Nutt. ex T. & G.
Smallflower Woodlandstar
Open woods and morainal slopes in the valley, vicinity of Two Ocean Lake. May–June.

*Mitella pentandra* Hook.
Five-star Mitrewort
Moist meadows and woods in the canyons. June–August.

*Mitella stauropetala* Piper
Smallflowered Mitrewort
Mesic sites in coniferous woods of the valley and canyons. May–August.

*Parnassia fimbriata* Konig
Rocky Mountain Parnassus
Stream banks and lakeshores from the valley to the subalpine. July–August.

* *Parnassia kotzebuei* Cham. ex Spreng.
Kotzebue Parnassus
Moist sites in the subalpine, Holly Lake Cirque. July–August.

*Parnassia palustris* L. var. *montanensis* (Fern. & Rydb.) Hitchc.
Marsh Parnassus
Moist sites in the subalpine, Holly Lake Cirque. July–August.

*Parnassia parviflora* DC.
Smallflowered Parnassus
Moist sites along the Snake River. June–August.

*Saxifraga adscendens* L. var. *oregonensis* (Raf.) Breit.
Ascending Saxifrage
Rocky slopes on alpine moraines. July–August.

*Saxifraga bronchialis* L. var. *austromontana* (Wieg.) Piper ex Jones
Yellowdot Saxifrage
Large boulders in the Hidden Falls area, also around subalpine lakes; circumboreal. July–August.

*Saxifraga caespitosa* L. var. *minima* Blank.
Tufted Saxifrage
Rocky areas in the alpine. July–August.

*Saxifraga cernua* L.
Nodding Saxifrage
Rocky slopes in the alpine, Sheep Mt. July–August.

*Saxifraga flagellaris* Willd. ex Sternb. var. *crandallii* (Gand.) Dorn
Stoloniferous Saxifrage
Rare on rock ledges of the Middle Teton and Grand Teton above 12,000 ft.; circumboreal. July–August.

*Saxifraga occidentalis* Wats.
Western Saxifrage
Moist sites in the canyons and mountain slopes. July–August.

*Saxifraga odontoloma* Piper
Brook Saxifrage
Stream banks in the canyons. July–August.

*Saxifraga oppositifolia* L.
Opposite-leaved Saxifrage
Rocky areas in the alpine; circumboreal. July–August.

*Saxifraga rhomboidea* Greene
Diamondleaf Saxifrage
Rock crevices or wet meadows in the subalpine and alpine. July–August.

*Saxifraga rivularis* L. ssp. *hyperborea* (R. Br.) Dorn
var. *debilis* (Engelm. ex Gray) Dorn
Pygmy Saxifrage
Rare on ledges above Amphitheater Lake. July–August.

*Saxifraga subapetala* E. Nels. [*S. oregana,* misapplied]
Oregon Saxifrage
Moist sites in the north end of the valley, Highway 89, 9 miles north of Jackson Lake Lodge. July–August.

*Telesonix heucheriformis* (Rydb.) Rydb. [*Boykinia heucheriformis* (Rydb.) Rosend.]
Telesonix
Limestone rock crevices, 1/3 mile south of Rendezvous Peak. July–August.

## Scrophulariaceae Figwort Family

*Besseya wyomingensis* (A. Nels.) Rydb.
Wyoming Kittentail
Moist open ridges and meadows in the subalpine and alpine, occasionally in the valley. July–August.

*Castilleja angustifolia* (Nutt.) G. Don var. *dubia* A. Nels. [*C. chromosa*]
Northwestern Paintbrush
Sagebrush flats in the valley. May–June.

*Castilleja cusickii* Greenm.
Cusick Paintbrush
Meadows and slopes along the Snake River. June–August.

*Castilleja flava* Wats.
Yellow Paintbrush
Dry soils frequently associated with sagebrush. June–August.

*Castilleja linariifolia* Benth.
Wyoming Paintbrush
Dry sagebrush flats of the valley; state flower of Wyoming, and the common hemiparasite of sagebrush. July–August.

*Castilleja miniata* Dougl. ex Hook.
Scarlet Paintbrush
Moist meadows, woods and slopes; from the valley to the subalpine. This taxon has 5 levels of polyploidy. July–August.

*Castilleja pilosa* (Wats.) Rydb. var. *longispica* (A. Nels.) Holmgren
Hairy Paintbrush
Morainal soils throughout the valley. July–August.

*Castilleja pulchella* Rydb.
Beautiful Paintbrush
Meadow and rocky slopes of the alpine, saddle between Middle and South Teton. July–August.

*Castilleja rhexifolia* Rydb.
Rhexia-leaved Paintbrush
Moist subalpine and alpine meadows and open woods, tends to prefer shade. July–August.

*Castilleja sulphurea* Rydb.
Sulphur Paintbrush
Moist subalpine meadows or exposed slopes. July–August.

*Collinsia parviflora* Lindl.
Blue-eyed Mary
Sagebrush flats in the valley. May–June.

*Cordylanthus ramosus* Nutt. ex Benth.
Bushy Birds Beak
Sagebrush flats in the valley, north end of Blacktail Butte. June–August.

*Limosella aquatica* L.
Mudwort
Shallow water or on wet mud in the valley. June–August.

† *Linaria dalmatica* (L.) Miller
Dalmatian Toadflax
Disturbed sites; native of Europe. July–August.

† *Linaria vulgaris* Miller
Butter-and-eggs
Disturbed sites; native of Europe. June–August.

* *Mimulus breweri* (Greene) Cov.
Brewer Monkey Flower
Cliffs, stream banks and slopes in the canyons. July–August.

*Mimulus floribundus* Lindl.
Floriferous Monkey Flower
Moist sites and stream banks in the canyons. June–July.

*Mimulus guttatus* DC.
Common Yellow Monkey Flower
Springs and stream banks in the valley and canyons. June–August.

*Mimulus lewisii* Pursh
Lewis Monkey Flower
Along streams in the canyons above 8,000 ft. July–August.

*Mimulus moschatus* Dougl. ex Lindl.
Musk Monkey Flower
Seeping streams, often in the shade; canyons. June–August.

*Mimulus suksdorfii* Gray
Suksdorf Monkey Flower
Valley to subalpine, mesic to exposed sites; Waterfalls Canyon. June–July.

*Mimulus tilingii* Regel
Subalpine Monkey Flower
Wet sites around springs and streams; this taxon is not always distinct from *M. guttatus*. July–August.

*Orthocarpus luteus* Nutt.
Yellow Owl-clover
Sagebrush flats and open woods along the Snake River. July–August.

*Orthocarpus tolmiei* H. & A.
Tolmie Owl-clover
Sagebrush flats in the valley. July–August.

*Pedicularis bracteosa* Benth. var. *paysoniana* (Pennell) Cronq.
Bracted Lousewort
Open coniferous forests from 6,700 ft. up to 8,500 ft. June–July.

*Pedicularis contorta* Benth. var. *contorta*
White-coiled Lousewort
Moist open coniferous forests from valley moraines to subalpine. July–August.

*Pedicularis cystopteridifolia* Rydb.
Fern-leaved Lousewort
Mountain slopes and meadows. July–August.

*Pedicularis groenlandica* Retz.
Elephanthead
Meadows and stream banks in the subalpine and alpine. July–August.

*Pedicularis parryi* Gray var. *parryi*; var. *purpurea* Parry
Parry Lousewort
Moist meadows and slopes in the subalpine and alpine. July–Sept.

*Pedicularis procera* Gray
Tall Lousewort; Gray Lousewort
Moist gravel along the Gros Ventre River. July–August.

*Pedicularis pulchella* Pennell
Beautiful Lousewort
Alpine and subalpine, Sheep Mt. July–August.

*Pedicularis racemosa* Dougl. ex Benth. var. *alba* (Pennell) Cronq.
Leafy Lousewort
Woods and meadows of the valley and extending into the canyons to 8,500 ft. July–August.

*Penstemon attenuatus* Dougl. ex Lindl. var. *pseudoprocerus* (Rydb.) Cronq.
Taper-leaved Penstemon
Open woods and meadows in the subalpine. July–August.

*Penstemon cyananthus* Hook.
Wasatch Penstemon
Disturbed gravel areas throughout the valley. June–July.

*Penstemon cyaneus* Pennell
Blueflowered Penstemon
Disturbed sites near Swan Lake (south of Colter Bay Village). June–July.

*Penstemon deustus* Dougl. ex Lindl.
Hot-rock Penstemon
Rocky slopes in the valley and extending into the canyons. June–July.

*Penstemon glaber* Pursh var. *glaber*
Rocky slopes, Gros Ventre Slide. July–August.

*Penstemon humilis* Nutt. ex Gray
Low Penstemon
Snake River bottoms in the south end of the valley. June–July.

*Penstemon montanus* Greene
Cordroot Penstemon
Rocky sites in the subalpine and alpine. July–August.

*Penstemon procerus* Dougl. ex Grah.
Small-flowered Penstemon
Moist to dry disturbed sites throughout the valley. June–August.

*Penstemon radicosus* A. Nels.
Matroot Penstemon
Hills and slopes in the valley, north end of Blacktail Butte. June–July.

*Penstemon rydbergii* A. Nels. var. *rydbergii*
Rydberg Penstemon
Meadows to stream banks, valley to subalpine. June–July.

*Penstemon subglaber* Rydb.
Smooth Penstemon
Exposed slopes and open woods in the valley. June–August.

*Penstemon whippleanus* Gray
Whipple Penstemon
Slopes and ridges in the subalpine and alpine; two distinct color phases—whitish and dark purple. July–August.

*Scrophularia lanceolata* Pursh
Lanceleaf Figwort
Rocky south facing slopes in the canyons. July–August.

*Synthyris pinnatifida* Wats.
Cutleaf Kittentails
Rocky slopes and ridges, subalpine to alpine. July–August.

† *Verbascum thapsus* L.
Common Mullein
Disturbed sites in the valley; native of Eurasia. July–August.

*Veronica americana* Schwein. ex Benth.
American Brooklime
Meadows, stream banks and springs in the valley. July–August.

† *Veronica anagallis-aquatica* L.
Water Speedwell
Along ditches and slow moving streams, Kelly Warm Spring; native of Eurasia. June–August.

*Veronica peregrina* L. var. *xalapensis* (H.B.K.) Pennell
Purslane Speedwell
Moist places or drying ponds, near Colter Bay. June–August.

* *Veronica scutellata* L.
Skullcap Speedwell
Ponds north of Colter Bay Village. June–August.

*Veronica serpyllifolia* L. var. *humifusa* (Dickson) Vahl.
Thyme-leaf Speedwell
Sedge meadows near Swan Lake. June–August.

*Veronica wormskjoldii* R. & S.
Alpine Speedwell
Moist places in subalpine and alpine. July–August.

## Solanaceae Nightshade Family

† *Hyoscyamus niger* L.
Black Henbane
Disturbed sites in the valley; contains alkaloids poisonous to man and livestock; native of Europe. July–Sept.

† *Solanum dulcamara* L.
European Bittersweet
Spalding Bay on Jackson Lake; berries are poisonous; native of Eurasia. June–August.

## Urticaceae Nettle Family

*Urtica dioica* L. ssp. *gracilis* (Aiton) Selander
Stinging Nettle
Moist sites from the Snake River up into the canyons. July–August.

## Valerianaceae Valerian Family

*Valeriana acutiloba* Rydb. var. *pubicarpa* (Rydb.) Cronq.
Cordilleran Valerian
Moist rocky slopes in the mountains. July–August.

*Valeriana edulis* Nutt. ex T. & G.
Taprooted Valerian
Meadows and open woods of the valley up into the subalpine. June–August.

*Valeriana occidentalis* Heller
Western Valerian
Moist, shaded sites in the subalpine, Phelps Lake Lookout and Skyline Trail. July–August.

## Verbenaceae Verbena Family

† *Verbena bracteata* Lag. & Rodr.
Prostrate Vervain
Dry disturbed sites in the valley; native American weed. May–Sept.

## Violaceae Violet Family

*Viola adunca* Smith
Blue Violet
Moist meadows from the valley to subalpine. June–July.

*Viola canadensis L.*
Canada White Violet
Moist woods and meadows in the valley. June–July.

*Viola macloskeyi* Lloyd var. *pallens* (Banks ex DC.) Hitchc.
Wild White Violet
Wet sites in open woods, exit at Bradley Lake. June–July.

*Viola nephrophylla* Greene
Bog Violet
Moist sites in the valley. June–July.

*Viola nuttallii* Pursh
Nuttall Violet
Sagebrush flats or open woods of the valley. May–June.

* *Viola orbiculata* Geyer ex Holz.
Darkwoods Violet
Stream banks and moist woods of the valley. June–July.

*Viola palustris* L.
Marsh Violet
Moist meadows and stream banks, south side of Leigh Lake. June–July.

*Viola purpurea* Kell. var. *venosa* (Wats.) Brain.
Goosefoot Violet
Many habitats; dry to moist sites, mostly in the valley. May–June.

*Viola renifolia* Gray var. *brainerdii* (Greene) Fern.
Kidney-leaved Violet
Woods and slopes, Teton Canyon. June–July.

### Viscaceae Mistletoe Family

*Arceuthobium americanum* Nutt. ex Engelm.
Lodgepole Pine Dwarf Mistletoe
Parasitic on Lodgepole Pine throughout the valley. May–June.

*Arceuthobium cyanocarpum* (A. Nels. ex Rydb.) A. Nels.
Limber Pine Dwarf Mistletoe
Not seen in Grand Teton National Park, but reported in nearby areas. June–July.

*Arceuthobium douglasii* Engelm.
Fir Dwarf Mistletoe
Usually parasitic on *Pseudotsuga* (Douglas-fir), rarely on *Abies* or *Picea*. Reported on Teton Pass Road. April–May.

## CLASS LILIOPSIDA MONOCOTS

### Alismataceae Water Plantain Family

*Sagittaria cuneata* Sheld.
Arrowhead
Ponds and slow moving streams, Two Ocean Lake and ponds in the vicinity of Jackson Lake. July–August.

### Cyperaceae Sedge Family

*Carex albonigra* Mack.
White and Black Sedge
Rocky slopes in the alpine. July–August.

*Carex aquatilis* Wahl.
Water Sedge
Wet sites in the valley, around the base of Signal Mt. June–August.

*Carex athrostachya* Olney
Slender-beaked Sedge
Wet sites and mud of ephemeral ponds, north of Signal Mt. June–August.

*Carex atrata* L.
Blackened Sedge
Moist places in the subalpine, Amphitheater Lake. July–Sept.

*Carex aurea* Nutt.
Golden Sedge
Wet sites along the Snake River. June–August.

*Carex backii* Boott
Back Sedge
Woods and thickets, Teton Canyon. June–August.

* *Carex buxbaumii* Wahl.
Buxbaum Sedge
Wet sites, shores of the valley lakes. June–August.

*Carex canescens* L.
Pale Sedge
Wet sites in canyons. June–August.

* *Carex cusickii* Mack. ex Piper & Beattie
Cusick Sedge
Wet sites, Moose Ponds south of Park Headquarters. June–August.

* *Carex diandra* Schrank
Lesser-panicled Sedge
Wet sites in the valley, Swan Lake. June–August.

*Carex disperma* Dewey
Softleaved Sedge
Wet sites, mostly in the shade, Cascade Canyon. June–August.

*Carex douglasii* Boott
Douglas Sedge
Ephemerally moist or wet sites in the valley and alpine, north of Signal Mt. June–August.

*Carex echinata* Murray [*C. muricata,* misapplied]
Rough Sedge
Boggy meadows, midmontane to subalpine. June–August.

*Carex elynoides* Holm
Kobresia Sedge
Krummholz and alpine communities. July–August.

*Carex filifolia* Nutt.
Threadleaf Sedge
Dry slopes above Elk Ranch Reservoir, south of Uhl Hill. June–July.

*Carex geyeri* Boott
Elk Sedge
Coniferous forests of the valley and canyons up to 9,500 ft. June–August.

*Carex haydeniana* Olney
Cloud Sedge
Moist or dry sites in the subalpine. June–August.

*Carex hoodii* Boott
Hood Sedge
Many plant communities in the canyons and subalpine. June–August.

*Carex illota* Bailey
Sheep Sedge
Wet sites, midmontane and subalpine. July–August.

*Carex interior* Bailey
Inland Sedge
Wet sites in canyons and subalpine, Teton Canyon. July–August.

* *Carex laeviculmis* Meinsh.
Smooth-stem Sedge
Wet sites in the canyons. June–August.

*Carex lanuginosa* Michx.
Woolly Sedge
Wet sites in the valley, north of Signal Mt. June–August.

*Carex lenticularis* Michx. var. *pallida* (Boott) Dorn
Lens Sedge
Wet sites in the canyons and subalpine. June–August.

* *Carex leptalea* Wahl.
Bristle-stalked Sedge
Wet sites in the valley. June–August.

* *Carex luzulina* Olney var. *ablata* (Bailey) Hermann
Woodrush Sedge
Woods and meadows in the subalpine and alpine. July–August.

*Carex microptera* Mack. var. *limnophila* (Hermann) Dorn
Small-wing Sedge
Wet sites north of Signal Mt. and the alpine. June–August.

*Carex nardina* Fries
Spikenard Sedge
Dry meadows and rocky places in the alpine. June–August.

*Carex nebrascensis* Dewey
Nebraska Sedge
Wet sites in the valley. June–August.

*Carex neurophora* Mack.
Alpine Nerved Sedge
Wet meadows and streams in the subalpine. July–August.

*Carex nigricans* Meyer
Black Alpine Sedge
Meadows and streams in the subalpine, shores of Amphitheater Lake and Holly Lake. July–August.

*Carex norvegica* Retz. var. *stevenii* (Holm) Dorn
Scandinavian Sedge
Wet places in the subalpine, Teton Canyon. June–August.

*Carex nova* Bailey
Black Sedge
Meadows and rocky sites in the alpine. July–August.

*Carex oederi* Retz. var. *recterostrata* (Bailey) Dorn
Green Sedge
Wet sites, shore of Bear Paw Lake. June–August.

*Carex paysonis* Clokey
Payson Sedge
Moist rocky slopes in subalpine and alpine. July–August.

*Carex petasata* Dewey
Liddon Sedge
Moist or dry sites, north of Signal Mt. June–August.

*Carex phaeocephala* Piper
Dunhead Sedge
Open coniferous woods in the subalpine and alpine. July–August.

*Carex praegracilis* Boott
Blackcreeper Sedge
Moist to wet sites in the valley. June–August.

*Carex praticola* Rydb.
Meadow Sedge
Woods in the subalpine, Grassy Lake Road. June–August.

*Carex pyrenaica* Wahl.
Pyrenean Sedge
Rocky slopes and meadows, Webb Canyon. July–August.

*Carex raynoldsii* Dewey
Raynolds Sedge
Many plant communities, along Snake River and Death Canyon. July–August.

*Carex rossii* Boott
Ross Sedge
Many communities from the valley to the subalpine. June–August.

*Carex rostrata* Stokes
Beaked Sedge
Wet places along the Snake River and shores of the valley lakes. June–August.

* *Carex sartwellii* Dewey
Sartwell Sedge
Moist areas. June–August.

*Carex stenophylla* Wahl.
Narrowleaved Sedge
Dry to moist sites east of Elk Ranch Reservoir. June–August.

*Carex vallicola* Dewey
Valley Sedge
Moist to dry slopes in sagebrush or aspen in the valley. June–August.

*Carex vesicaria* L.
Blister Sedge
Wet sites, north of Signal Mt. July–August.

*Eleocharis acicularis* (L.) R. & S.
Needle Spikerush
Wet sites, often mudflats, Two Ocean Lake. June–Sept.

*Eleocharis flavescens* (Poir.) Urban var. *thermalis* (Rydb.) Cronq.
Yellow Spikerush
Wet areas, usually near hot springs. June–August.

*Eleocharis palustris* (L.) R. & S.
Creeping Spikerush
Wet sites north of Signal Mt., also shores of String Lake. June–August.

*Eleocharis quinqueflora* (Hartm.) Schw. [*E. pauciflora*]
Fewflowered Spikerush
Wet sites north of Signal Mt. and the subalpine. June–August.

*Eleocharis rostellata* (Torr.) Torr.
Beaked Spikerush
Wet places, Huckleberry Hot Springs. June–August.

* *Eriophorum gracile* Koch
Slender Cotton-grass
Sedge meadow or shallow water, south side of Leigh Lake. July–August.

*Eriophorum polystachium* L.
Tall Cotton-grass
Wet sites in the subalpine. July–August.

*Eriophorum viridicarinatum* (Engelm.) Fern.
Green-keeled Cotton-grass
Wet sites in the mountains, one mile below Columbine Cascade in Waterfalls Canyon. July–August.

*Scirpus acutus* Muhl. ex Bigelow
Hardstem Bulrush
Shallow ponds in the north end of the valley. June–August.

*Scirpus americanus* Pers.
Olney Threesquare
Wet sites in the valley. June–August.

## Hydrocharitaceae Frogbit Family

*Elodea canadensis* Michx.
Canada Waterweed
Floating in water, ponds and lakes; appears sporadically. June–Sept.

*Elodea longivaginata* St. John
Long-sheath Waterweed
Floating in water, ponds and lakes; appears sporadically. June–Sept.

## Iridaceae Iris Family

*Iris missouriensis* Nutt.
Missouri Iris
Moist meadows and stream banks, west side of Teton Range. June–July.

*Sisyrinchium idahoense* Bicknell var. *occidentale* (Bicknell) Henderson
Idaho Blue-eyed Grass
Marshes and wet meadows along the Snake River. June–August.

## Juncaceae Rush Family

*Juncus balticus* Willd. var. *montanus* Engelm. [*J. arcticus* Willd.]
Wiregrass
Moist sites along the Snake River. May–July.

*Juncus bufonius* L. var. *bufonius*
Toad Rush
Moist sites along the Snake River, nearly cosmopolitan. May–August.

*Juncus confusus* Cov.
Colorado Rush
Dry and wet meadows along the Snake River. June–August.

*Juncus drummondii* Meyer
Drummond Rush
Meadows and wooded slopes in the subalpine and alpine. June–August.

*Juncus ensifolius* Wikst. var. *montanus* (Engelm.) Hitchc.
Swordleaf Rush
Wet sites along the Snake River. June–August.

* *Juncus filiformis* L.
Thread Rush
Wet sites along the Snake River. June–August.

*Juncus longistylis* Torrey
Longstyle Rush
Wet sites in the valley and possibly into the subalpine. June–August.

*Juncus mertensianus* Bong.
Mertens Rush
Stream banks and lakeshores above 9,000 ft. June–August.

*Juncus nevadensis* Wats.
Nevada Rush
Moist areas, valley to subalpine. June–August.

*Juncus nodosus* L.
Jointed Rush
Along streams and rivers, southwest of Jackson. July–August.

*Juncus parryi* Engelm.
Parry Rush
Wet meadows and subalpine slopes. June–August.

*Juncus torreyi* Cov.
Torreys Rush
Moist sites in the valley; not seen since 1895. June–July.

*Juncus tracyi* Rydb.
Tracys Rush
Moist sites in the mountains, east face of Symmetry Spire. July–August.

* *Luzula glabrata* (Hoppe ex Rostk.) Desv. var. *hitchcockii* (Hämet-Ahti) Dorn
Smooth Woodrush
Meadows and woods in the subalpine, Holly Lake cirque. July–August.

*Luzula parviflora* (Ehrh.) Desv.
Millet Woodrush
Moist and wet sites, subalpine; circumboreal. June–August.

*Luzula spicata* (L.) DC.
Spike Woodrush
Rocky sites in subalpine and alpine; circumboreal; near summit of Table Mt. July–August.

*Luzula wahlenbergii* Rupr.
Wahlenberg Woodrush
Moist sites in the subalpine and alpine. July–August.

## Juncaginaceae Arrowgrass Family

*Triglochin maritimum* L. var. *elatum* (Nutt.) Gray.
Shore Arrowgrass
Wet alkaline soil, Huckleberry Hot Springs in the J.D.R. Parkway; circumboreal; poisonous to livestock. June–August.

*Triglochin palustre* L.
Swamp Arrowgrass
Wet meadows in wooded areas, south of Jackson; circumboreal. July–August.

## Lemnaceae   Duckweed Family

*Lemna minor* L.
Lesser Duckweed
Floating aquatic in ponds in the vicinity of Jackson Lake Lodge; cosmopolitan. Rarely flowers.

*Lemna trisulca* L.
Ivy-leaf Duckweed
Floating aquatic in streams, ponds and lakes in the valley; circumboreal. Rarely flowers.

* *Lemna valdiviana* Philippi
Valdivia Duckweed
Floating aquatic in sluggish streams and ponds in the valley, Kelly Warm Spring. Rarely flowers.

* *Spirodela polyrhiza* (L.) Schleiden
Great Duckweed
Ponds and sluggish streams in the valley. Rarely flowers.

## Liliaceae   Lily Family

*Allium acuminatum* Hook.
Tapertip Onion
Dry open morainal slopes near Moose. June–July.

*Allium brevistylum* Wats.
Shortsyle Onion
Riparian sites along the Snake River. June–July.

*Allium cernuum* Roth
Nodding Onion
Open wooded areas in canyons. July–August.

*Allium geyeri* Wats. var. *tenerum* Jones
Geyer Onion
Gravelly sites below Jackson Lake Dam; this variety has many bulbils replacing the flowers. June–July.

*Allium schoenoprasum* L.
Chive
Along the Snake River and into some canyons; circumboreal. June–August.

† *Asparagus officinalis* L.
Garden Asparagus
Cultivated food plant escaping to stream banks; native of Eurasia. June.

*Calochortus eurycarpus* Wats.
Big-pod Mariposa; Sego Lily
Sagebrush areas, Gros Ventre Slide Road. June–July.

*Calochortus nuttallii* T. & G.
Nuttalls Sego Lily
Foothills and canyons in the sagebrush. June.

*Camassia quamash* (Pursh) Greene
Blue Camas
Moist meadows in the valley north of Jackson Lake Dam; bulbs were eaten by Native Americans. June.

*Disporum trachycarpum* (Wats.) Benth. & Hook.
Fairy Bells
Shady sites near streams, along Snake River and in Death Canyon. June.

*Erythronium grandiflorum* Pursh
Glacier Lily
Meadows and open woods in the canyons and subalpine. June–August, depending on melting snowbanks.

*Fritillaria atropurpurea* Nutt.
Leopard Lily
Dry to mesic sites on moraines of the valley. June–July.

*Fritillaria pudica* (Pursh) Spreng.
Yellow Bell
Sagebrush flats of the valley; frequently reproduces asexually by bulblets. April–June.

*Lloydia serotina* (L.) Reichenb.
Alplily
Rocky sites in the alpine. July–August.

*Maianthemum racemosum* (L.) Link var. *amplexicaule* (Nutt.) Dorn [*Smilacina r.*]
False Solomon Seal
Moist sites in the shade, Snake River and shores of the valley lakes. June.

*Maianthemum stellatum* (L.) Link [*Smilacina s.*]
Starry False Solomon Seal
Moist sites in the shade, valley to midmontane. June.

*Streptopus amplexifolius* (L.) DC.
Clasping Twisted Stalk
Stream banks in the valley and lower portions of the canyons. June–July.

*Tofieldia glutinosa* (Michx.) Pers. var. *montana* (Hitchc.) Davis
False Asphodel
Shores of the valley lakes, also some subalpine lakes above 9,000 ft. June–August.

*Triteleia grandiflora* Lindl.
Wild Hyacinth
Meadows or open woods in vicinity of Sawmill ponds and Park Headquarters. June.

*Veratrum californicum* Durand
False Hellebore
Wet meadows in the vicinity of Phelps Lake and the north boundary of the park on Highway 89. Entire plant contains alkaloids that poison the cardiovascular system in mammals. July–August.

*Xerophyllum tenax* (Pursh) Nutt.
Beargrass
Open woods at the east end of Grassy Lake Reservoir, also in the 1974 burn area in the mouth of Waterfalls Canyon. July–August.

*Zigadenus elegans* Pursh
Mountain Death Camas
Rocky slopes and shores of subalpine lakes; poisonous to livestock and humans. July–August.

*Zigadenus paniculatus* (Nutt.) Wats.
Foothill Death Camas
Sagebrush and open woods in the valley; poisonous to livestock and humans. May–June.

*Zigadenus venenosus* Wats. var. *gramineus* (Rydb.) Walsh ex Peck
Meadow Death Camas
Moist and drying meadows in the canyons to subalpine; poisonous to livestock and humans. July–August.

## Najadaceae Water Nymph Family

* *Najas guadalupensis* (Spreng.) Morong
Southern Waternymph
Slow moving streams, ponds or lakes. June–August.

## Orchidaceae Orchid Family

*Calypso bulbosa* (L.) Oakes
Fairy Slipper
Shade of coniferous woods in the valley, area below Hidden Falls and south of Cattleman's Bridge; mycotrophic. June.

*Corallorhiza maculata* Raf.
Spotted Coralroot
Shade of coniferous woods around the shores of the valley lakes; mycotrophic. June–July.

* *Corallorhiza mertensiana* Bong.
Western Coralroot
Shade of coniferous woods around the shores of the valley lakes; mycotrophic. June–July.

*Corallorhiza striata* Lindl.
Striped Coralroot
Coniferous woods in the valley; mycotrophic. June–July.

*Corallorhiza trifida* Chat.
Early Coralroot
Rare, shade of coniferous woods, west shore of Leigh and Jackson Lake. July–August.

*Corallorhiza wisteriana* Conrad
Wister Coralroot
Shade of coniferous woods, Nature Trail Island at Colter Bay; mycotrophic. June–July.

* *Epipactis gigantea* Dougl. ex Hook.
Giant Helleborine
Wet shady areas along streams, east of Kelly. July.

*Goodyera oblongifolia* Raf.
Rattlesnake Plantain
Coniferous woods, String Lake and second switchback above Hidden Falls Bridge. July–August.

*Habenaria dilatata* (Pursh) Hook.
White Bog Orchid
Wet stream banks and boggy sites in the valley and canyons. June–August.

*Habenaria hyperborea* (L.) R. Br. [*Platanthera hyperborea*]
Northern Bog Orchid
Wet sites along the Snake River. June–August.

*Habenaria obtusata* (Banks ex Pursh) Richardson
Small Bog Orchid
Stream banks and wet meadows, north side Signal Mt. and Webb Canyon. June–August.

*Habenaria saccata* Greene
Slender Bog Orchid
Wet boggy areas, Granite Creek. June–August.

*Habenaria unalascensis* (Spreng.) Wats. [*Piperia unalascensis*]
Alaska Bog Orchid
Dry woods to moist sites, east shore of Jackson Lake. June–August.

*Listera borealis* Morong
Northern Twayblade
Wet shady sites in spruce-fir forest. July–August.

* *Listera caurina* Piper
Western Twayblade
Moist sites in coniferous woods, Webb Canyon and White Grass Ranger Station. July–August.

* *Listera convallarioides* (Sw.) Nutt. ex Ell.
Broad-leaved Twayblade
Shade of conifers, west shore of Jenny Lake. June–July.

*Listera cordata* (L.) R. Br.
Heart-leaved Twayblade
Moist slopes in shade of conifers, Elk Island and Indian Paintbrush Canyon. June–July.

*Spiranthes romanzoffiana* Cham.
Hooded Ladies Tresses
Wet sites in coniferous woods, Snake River and ponds near Colter Bay. July–August.

## Poaceae Grass Family

† *Agropyron cristatum* (L.) Gaertn. var. *cristatum*
Crested Wheatgrass
Along roads and other disturbed sites; native of Eurasia. June–August.

*Agrostis exarata* Trin.
Spike Redtop
Moist open areas. July–August.

*Agrostis humilis* Vasey
Alpine Bentgrass
Mostly alpine or subalpine. July–August.

*Agrostis idahoensis* Nash
Idaho Bentgrass
Mountain meadows. July–August.

* *Agrostis oregonensis* Vasey
Oregon Bentgrass
Wet areas in the valley. July–August.

*Agrostis scabra* Willd.
Ticklegrass
Moist areas in the valley. July–August.

† *Agrostis stolonifera* L. [*A. alba,* misapplied]
Redtop
Moist areas in the valley; native of Europe. June–August.

*Agrostis thurberiana* Hitchc.
Thurber Bentgrass
Moist areas in the mountains. July–August.

*Agrostis variabilis* Rydb.
Variable Bentgrass
Moist areas in the mountains. July–August.

*Alopecurus aequalis* Sobol.
Short-awned Foxtail
In water and wet areas in the valley and subalpine. June–August.

*Alopecurus alpinus* Smith
Smooth Alpine Foxtail
Meadows and stream banks in the mountains. June–August.

† *Avena fatua* L.
Wild Oats
Weed in waste areas; native of Europe. June–August.

*Beckmania syzigachne* (Steudel) Fern.
American Sloughgrass
Wet marshy areas, pond north of Brinkerhoff Cabin. July–August.

*Bromus anomalus* Rupr. ex Fourn.
Nodding Brome
Dry to moist sites in meadows and open woods. June–August.

*Bromus carinatus* H. & A.
California Brome
Moist habitats from the valley to subalpine forest. June–August.

*Bromus ciliatus* L.
Fringed Brome
Moist wood habitats in the valley. July–August.

† *Bromus commutatus* Schrad.
Hairy Chess
Dry to moist disturbed sites in the valley; native of Europe. June–July.

† *Bromus inermis* Leyss.
Smooth or Common Brome
Disturbed sites, woods and meadows; native of Eurasia. June–August.

† *Bromus japonicus* Thunb. ex Murray
Japanese Chess
Disturbed sites in the valley; native of Eurasia. June–July.

† *Bromus tectorum* L.
Downy Chess; Cheatgrass
Disturbed sites in the sagebrush zone; native of Eurasia. May–July.

*Bromus vulgaris* (Hook.) Shear
Common Brome
Woods in the valley. June–August.

*Calamagrostis canadensis* (Michx.) Beauv.
Canada Reedgrass
Wet sites in the mountains. July–August.

*Calamagrostis inexpansa* Gray
Narrow-spiked Reedgrass
Moist sites. July–August.

* *Calamagrostis koelerioides* Vasey
Fire Reedgrass
Mountain meadows and slopes. July–August.

*Calamagrostis purpurascens* R. Br.
Purple Reedgrass
Rocky sites in the mountains. July–August.

*Calamagrostis rubescens* Buckl.
Pinegrass
Dry to moist sites in lower elevations in the mountains. July–August.

* *Calamagrostis scopulorum* Jones
Cliff Reedgrass
Moist areas. July–August.

*Calamagrostis stricta* (Timm) Koeler
Slimstem or Northern Reedgrass
Moist to wet sites in the valley. July–August.

*Catabrosa aquatica* (L.) Beauv.
Brookgrass
Shallow water or wet places. June–August.

*Cenchrus longispinus* (Hack.) Fern.
Sandbur
Weed in sandy sites, Snake River Canyon. July–August.

*Cinna latifolia* (Trev.) Griseb.
Drooping Woodreed
Moist woods and stream banks in lower montane region. July–August.

† *Dactylis glomerata* L.
Orchard Grass
Meadows and disturbed sites in the valley; native of Europe. June–August.

*Danthonia californica* Bolander
California Oatgrass
Meadows and open woods. June–August.

*Danthonia intermedia* Vasey
Timber Oatgrass
Dry to moist slopes in the woods up to subalpine. July–August.

*Danthonia spicata* (L.) Beauv. ex R. & S.
Poverty Oatgrass
Woods and open areas. July–August.

*Danthonia unispicata* (Thurb.) Munro ex Macoun
One-spike Oatgrass
Sagebrush flats of the valley. June–August.

*Deschampsia atropurpurea* (Wahl.) Scheele
Mountain Hairgrass
Mountain meadows and canyons. July–August.

*Deschampsia cespitosa* (L.) Beauv.
Tufted Hairgrass
Wet areas in the subalpine and alpine. July–August.

*Deschampsia danthonioides* (Trin.) Munro
Annual Hairgrass
Wet or seasonally wet areas. June–August.

* *Deschampsia elongata* (Hook.) Munro
Slender Hairgrass
Wet shady areas in the mountains. July–August.

*Distichlis stricta* (Torr.) Rydb.
Saltgrass
Alkaline areas east of Elk Ranch Reservoir. June–August.

*Elymus canadensis* L.
Canada Wildrye
Open sites especially where moist. June–August.

*Elymus cinereus* Scribn. & Merr. [*Leymus cinereus*]
Great Basin Rye
Sagebrush meadows and roadsides. June–August.

*Elymus elymoides* (Raf.) Swezey [*Sitanion hystrix*]
Squirreltail
Dry sagebrush to alpine. June–August.

*Elymus glaucus* Buckl.
Blue Wildrye
Open woods and hills at middle elevations. July–August.

† *Elymus hispidus* (Opiz) Melderis [*Agropyron intermedium*]
Intermediate Wheatgrass
Disturbed areas; native of Europe. June–August.

*Elymus lanceolatus* (Scribn. & Sm.) Gould
var. *lanceolatus* [*Agropyron dasystachyum*]
Thickspike Wheatgrass
Sagebrush or open areas; both awned and unawned forms occur. June–August.

† *Elymus repens* (L.) Gould [*Agropyron repens*]
Quackgrass
Widespread weed in disturbed sites; native of Eurasia. June–August.

*Elymus scribneri* (Vasey) Jones [*Agropyron scribneri*]
Scribner Wheatgrass
Rocky ridges in subalpine and alpine. July–August.

*Elymus smithii* (Rydb.) Gould [*Agropyron smithii*]
Western Wheatgrass
Outwash plains and moraines in the valley. June–August.

*Elymus spicatus* (Pursh) Gould [*Agropyron spicatum*]
Bluebunch Wheatgrass
Hills and slopes in the valley. June–August.

*Elymus trachycaulus* (Link) Gould [*Agropyron caninum,* misapplied]
ssp. *trachycaulus* var. *trachycaulus* Slender Wheatgrass;
var. *andinus* (Scribn. & Sm.) Dorn [*Agropyron subsecundum*]
Bearded Wheatgrass
Dry to moist habitats, from the valley to subalpine. A variable species needing detailed study. June–August.

*Festuca brachyphylla* Schultes & Schultes [*F. ovina* var. *brevifolia*]
Alpine Fescue
Meadows in the subalpine and alpine. July–August.

*Festuca idahoensis* Elmer
Idaho Fescue
Dry sagebrush flats to subalpine meadows. June–August.

*Festuca rubra* L.
Red Fescue
Moist meadows and open woods, from the valley to subalpine. July–August.

*Festuca saximontana* Rydb. [*F. ovina* var. *rydbergii*]
Sheep Fescue
Montane to alpine. July–August.

*Festuca subulata* Trin.
Bearded Fescue
Moist shady areas, from the valley to midmontane. June–August.

*Glyceria borealis* (Nash) Batchelder
Northern Mannagrass
Moist areas in the valley. June–August.

*Glyceria elata* (Nash ex Rydb.) Jones
Tall Mannagrass
Moist areas in the valley. June–August.

*Glyceria grandis* Wats.
American Mannagrass
Wet areas in the valley. June–August.

*Glyceria striata* (Lam.) Hitchc.
Fowl Mannagrass
Along slow streams in wooded areas. July–August.

*Hierochloe odorata* (L.) Beauv.
Sweetgrass
Moist areas from the valley to the subalpine. June–August.

*Hordeum brachyantherum* Nevski
Meadow Barley
Open, mostly moist areas. June–August.

*Hordeum jubatum* L.
Foxtail Barley
Roadsides and other disturbed sites. June–August.

*Koeleria macrantha* (Ledeb.) Schultes
Junegrass
Sagebrush flats and open woods. June–August.

*Leucopoa kingii* (Wats.) Weber
Western Grass
Slopes and open woods in the valley. June–August.

*Melica bulbosa* Geyer ex Porter & Coult.
Oniongrass
Dry to moderately moist slopes in the valley. June–August.

* *Melica smithii* (Porter ex Gray) Vasey
Smith Melicgrass
Moist woods in the valley. June–August.

*Melica spectabilis* Scribn.
Purple Oniongrass
Meadows and woods in the valley. June–August.

*Melica subulata* (Griseb.) Scribn.
Alaska Oniongrass
Meadows and shaded areas. June–August.

*Muhlenbergia andina* (Nutt.) Hitchc.
Foxtail Muhly
Moist areas in the valley. July–August.

*Muhlenbergia asperifolia* (Nees & Meyer) Parodi
Scratchgrass Muhly
Moist, usually alkaline areas in the valley. July–August.

*Muhlenbergia filiformis* (Thurb.) Rydb.
Pull-up Muhly
Open woods and meadows. June–August.

*Muhlenbergia racemosa* (Michx.) B.S.P.
Satin-grass
Moist or dry areas in the valley. June–August.

*Muhlenbergia richardsonis* (Trin.) Rydb.
Mat Muhly
Moist or dry sites in the valley. July–August.

*Nassella viridula* (Trin.) Barkw. [*Stipa viridula*]
Green Needlegrass
Sagebrush flats. June–August.

*Oryzopsis asperifolia* Michx.
Roughleaf Ricegrass
Woods and slopes in the valley. June–August.

*Oryzopsis exigua* Thurb.
Little Ricegrass
Wooded slopes usually under conifers. June–August.

*Panicum capillare* L.
Witchgrass
Moist open sites along ditches and streams. June–August.

*Phalaris arundinacea* L.
Reed Canary Grass
Moist disturbed sites; circumboreal. June–August.

† *Phalaris canariensis* L.
Common Canary Grass
Disturbed sites; native of Europe. June-August.

*Phleum alpinum* L.
Alpine Timothy
Meadows in the subalpine and alpine. July–August.

† *Phleum pratense* L.
Common Timothy
Meadows and disturbed sites in the valley; native of Eurasia. June–August.

*Phragmites australis* (Cav.) Trin.
Common Reed
Marshes and shallow water, north end of String Lake; most widely distributed species of all angiosperms. July–August.

*Poa alpina* L.
Alpine Bluegrass
Mountain meadows or talus. July–August.

† *Poa annua* L.
Annual Bluegrass
Weed of disturbed areas, open woods and meadows; native of Eurasia. June–August.

*Poa arctica* R. Br. var. *grayana* (Vasey) Dorn
Arctic Bluegrass
Moist meadows and open ridges in the alpine. July–August.

*Poa arida* Vasey
Plains Bluegrass
Sagebrush in the valley to subalpine. June–August.

† *Poa bulbosa* L.
Bulbous Bluegrass
Disturbed areas and meadows in the valley; native of Eurasia and Africa. June–July.

† *Poa compressa* L.
Canada Bluegrass
Disturbed areas and meadows in the valley; native of Eurasia. June–August.

* *Poa curta* Rydb.
Shortleaf Bluegrass
Moist shaded areas in the valley. June–July.

*Poa cusickii* Vasey var. *cusickii*; var. *epilis* (Scribn.) Hitchc.
Cusick Bluegrass
Sagebrush flats and meadows in the valley. May–August.

*Poa fendleriana* (Steudel) Vasey
Mutton Grass
Meadows and slopes in the valley. June–August.

*Poa interior* Rydb.
Inland Bluegrass
Open slopes and woods in the valley. June–August.

*Poa leptocoma* Trin.
Bog Bluegrass
Stream banks and meadows in the subalpine and alpine. July–August.

*Poa nervosa* (Hook.) Vasey var. *wheeleri* (Vasey) Hitchc.
Wheeler Bluegrass
Open woods and slopes in the valley. June–August.

*Poa palustris* L.
Fowl Bluegrass
Moist sites in the valley. June–August.

*Poa pattersonii* Vasey
Patterson Bluegrass
Alpine or subalpine. July–August.

*Poa pratensis* L.
Kentucky Bluegrass
Moist to moderately dry areas; cosmopolitan. June–August.

*Poa reflexa* Vasey & Scribn.
Nodding Bluegrass
Wet sites on stream banks in the alpine or subalpine. July–August.

*Poa rupicola* Nash ex Rydb.
Timberline Bluegrass
Rocky slopes in the alpine. July–August.

*Poa secunda* Presl
Sanberg Bluegrass
Valley, subalpine and alpine. Kellogg (1985) has shown that this species is highly variable. Plants growing in different habitats look so different that they have traditionally been treated as distinct species. *Poa ampla* Merr., *P. canbyi* (Scribn.) Howell, *P. gracillima* Vasey, *P. juncifolia* Scribn., *P. nevadensis* Vasey, *P. sanbergii* Vasey, and *P. scabrella* (Thurb.) Benth. are among the species that Kellogg showed to be part of *P. secunda.* May–July.

† *Polypogon monspeliensis* (L.) Desf.
Rabbitfoot Grass
Moist soil or in water; native of Eurasia and Africa. June–August.

* *Puccinellia fernaldii* (Hitchc.) Voss
Fernald Alkaligrass
Moist areas or in water. June–August.

*Puccinellia nuttalliana* (Schultes) Hitchc.
Nuttall Alkaligrass
Moist, usually alkaline sites, east of Elk Ranch Reservoir. June–August.

*Puccinellia pauciflora* (Presl) Munz var. *holmii* (Beal) Hitchc.
Weak Alkaligrass
Moist areas or in water. June–August.

*Schizachne purpurascens* (Torr.) Swallen
False Melic
Moderately moist woods and open hillsides. June–August.

† *Setaria viridis* (L.) Beauv.
Green Bristlegrass
Weed of disturbed sites; native of Eurasia. June–August.

*Spartina gracilis* Trin.
Alkali Cordgrass
Moist alkaline areas. June–August.

*Sphenopholis obtusata* (Michx.) Scribn. var. *major* (Torr.) Erdman
Wedgegrass
Moist to wet areas. July–August.

*Stipa comata* Trin. & Rupr.
Needle & Thread Grass
Sagebrush flats in the valley. June–July.

*Stipa hymenoides* Roemer & Schultes
[*Oryzopsis hymenoides* (Roemer & Schultes) Ricker]
Indian Ricegrass
Sandy to well drained soils, from the valley to midmontane. June–August.

*Stipa lettermanii* Vasey
Letterman Needlegrass
Sagebrush slopes and open woods. July–August.

*Stipa nelsonii* Scribn. var. *nelsonii*
[*S. williamsii, S. columbiana* sensu Hitchc., non Macoun]
Columbia Needlegrass
Sagebrush flats and slopes. June–August.

*Stipa occidentalis* Thurb. ex Wats.
Small Needlegrass
Sagebrush flats and open woods. June–August.

*Stipa richardsonii* Link
Richardson Needlegrass
Woods and meadows. June–August.

*Trisetum spicatum* (L.) Richt.
Spike Trisetum
Subalpine woods and alpine slopes. July–August.

*Trisetum wolfii* Vasey
Beardless Trisetum
Sagebrush flats and meadows in the valley. June–July.

## Potamogetonaceae Pondweed Family

*Potamogeton alpinus* Balbis
Northern Pondweed
Ponds and lakes. June–August.

*Potamogeton amplifolius* Tuckerman
Large-leaved Pondweed
Ponds and lakes. July–August.

*Potamogeton epihydrus* Raf.
Ribbon-leaf Pondweed
Slow moving streams, ponds and lakes. June–July.

*Potamogeton filiformis* Pers.
Filiform Pondweed
Ponds in the north end of the valley. June–August.

*Potamogeton foliosus* Raf.
Leafy Pondweed
Ponds, lakes and slow moving streams. July–August.

* *Potamogeton friesii* Rupr.
Fries Pondweed
Ponds and lakes. July–August.

*Potamogeton gramineus* L.
Grass Pondweed
Ponds and lakes. June–August.

*Potamogeton natans* L.
Floating Pondweed
Ponds and lakes, Swan Lake north of Colter Bay. June–August.

*Potamogeton obtusifolius* Mertens & Koch
Blunt-leaved Pondweed
Ponds and lakes. July–August.

*Potamogeton pectinatus* L.
Fennel-leaf Pondweed
Ponds and lakes. June–July.

*Potamogeton pusillus* L.
Dwarf Pondweed
Ponds, lakes and shallow streams. July–August.

*Potamogeton richardsonii* (Bennett) Rydb.
Richardson Pondweed
Ponds and lakes. June–August.

* *Potamogeton zosteriformis* Fern.
Eelgrass Pondweed
Ponds and lakes, Swan Lake north of Colter Bay. June–August.

## Sparganiaceae Bur-reed Family

*Sparganium angustifolium* Michx.
Narrowleaved Bur-reed
Beaver ponds along the Snake River and ponds near Colter Bay. July.

*Sparganium emersum* Rehm.
Emersed Bur-reed
Shallow ponds, wet meadows or lakes. July.

## Typhaceae Cattail Family

*Typha angustifolia* L.
Narrow-leaved Cattail
Slow moving water and ponds. Blooming spikes and rhizomes are edible. June–August.

*Typha latifolia* L.
Common Cattail
Slow moving water, ponds east of Jackson Lake. Blooming spikes and rhizomes are edible. June–August.

## Zannichelliaceae Horned Pondweed Family

*Zannichellia palustris* L.
Horned Pondweed
Aquatic. June–August.

## Statistical Summary

| | Indigenous Family | Indigenous Genera | Indigenous Species | Alien Family | Alien Genera | Alien Species |
|---|---|---|---|---|---|---|
| Fern & Fern Allies | 7 | 19 | 32 | 0 | 0 | 0 |
| Gymnosperms | 2 | 4 | 9 | 0 | 1 | 1 |
| Dicots | 64 | 256 | 653 | 7 | 48 | 98 |
| Monocots | 15 | 75 | 245 | 0 | 4 | 18 |
| | 88 | 354 | 939 | 7 | 53 | 117 |

### Largest Families
(Native + Alien Species)

| | Native | | Alien |
|---|---|---|---|
| Asteraceae | 142 | + | 24 |
| Poaceae | 101 | + | 17 |
| Brassicaceae | 40 | + | 18 |
| Cyperaceae | 55 | + | 0 |
| Scrophulariaceae | 49 | + | 4 |
| Fabaceae | 30 | + | 10 |
| Rosaceae | 38 | + | 1 |

### Grand Totals

| | |
|---|---|
| Families | 95 |
| Genera | 407 |
| Species | 1056 |

### Largest Genera
(Native + Alien Species)

| | Native | | Alien |
|---|---|---|---|
| Carex | 46 | + | 0 |
| Salix | 20 | + | 0 |
| Poa | 15 | + | 3 |
| Erigeron | 17 | + | 0 |
| Ranunculus | 15 | + | 1 |
| Artemisia | 13 | + | 1 |
| Potentilla | 14 | + | 0 |
| Senecio | 14 | + | 0 |
| Juncus | 13 | + | 0 |
| Potamogeton | 13 | + | 1 |

## Poisonous Species

| | |
|---|---|
| *Pteridium aquilinum* | Bracken Fern |
| *Hyoscyamus niger* | Black Henbane |
| *Solanum dulcamara* | European Bittersweet |
| *Prunus virginianum* | Chokecherry |
| *Arnica cordifolia* | Heartleaf Arnica |
| *Sambucus racemosa* | Elderberry |
| *Apocynum androsaemifolium* | Spreading Dogbane |
| *Apocynum cannabinum* | Hemp Dogbane |
| *Chenopodium album* | Lambquarters |
| *Tanacetum vulgare* | Common Tansy |
| *Rheum rhubarbarum* | Rhubarb |
| *Aconitum columbianum* | Columbia Monkshood |
| *Delphinium nuttallianum* | Nuttall Larkspur |
| *Delphinium occidentale* | Duncecap Larkspur |
| *Cicuta maculata* | Water Hemlock |
| *Conium maculatum* | Poison Hemlock |
| *Astragalus bisulcatus* | Two-grooved Milkvetch |
| *Astragalus canadensis* | Canada Milkvetch |
| *Astragalus cibarius* | Browse Milkvetch |
| *Astragalus miser* | Weedy Milkvetch |
| *Lupinus argenteus* | Silver Lupine |
| *Lupinus sericeus* | Silky Lupine |
| *Actaea rubra* | Western Baneberry |
| *Xanthium strumarium* | Cocklebur |
| *Toxicodendron rydbergii* | Poison Ivy |
| *Triglochin maritimum* | Shore Arrowgrass |
| *Veratrum californicum* | False Hellebore |
| *Zigadenus elegans* | Mountain Death Camas |
| *Zigadenus paniculatus* | Foothill Death Camas |
| *Zigadenus venenosus* | Meadow Death Camas |

## Plant Species of Concern, Grand Teton National Park

The National Park Service operates within a policy of protecting all native species. Therefore, special designation for species and lands are not needed to maintain species of concern. In actuality though, specific recognition and monitoring of species is often required to avoid visitor over-use which could easily impact on species which are geographically disjunct from their overall ranges and endemics (species restricted to a very limited area globally). In order to pursue such a policy, Grand Teton National Park began a project in 1989 with the Wyoming Natural Diversity Base (WYNDD) of the Nature Conservancy. Hollis Marriott, as director of the project, has assembled a list of plant species of concern which contains 66 taxa. The asterisk "*" precedes all species of concern in the species listing. For criteria used, future monitoring plans and management recommendations see *Rare Plants of GTNP—Second Annual Report—1991.* It should be emphasized that the list of 66 species in this report is still preliminary and will be revised as additional information becomes available through field surveys conducted by the WYNDD.

## Recommendations to Maintain the Species Diversity of Grand Teton National Park Flora for the Future

1. Continue monitoring the species of concern project with the Nature Conservancy and then use this information in vegetation management.
2. Include plant ecologists in any planning for new road and parking area construction in the park. Consider a moratorium on all road and parking lot construction. Such projects create the disturbance sites for the quick invasions of alien species. If visitation continues to increase, alternative "people movers" plans should be developed (busses and open trams).
3. Place stricter controls on horseback use in the back country. The transportation of feed supplies to back country campsites represents a threat to pristine areas. Increased horse travel equals more trail disturbance and more alien invasion.
4. Resource management plans should involve the following:
   a. A continued collection of plants in all areas of the park by qualified plant taxonomists who will place their specimens in the park herbarium.
   b. Any revegetation plans should involve a qualified plant taxonomist to check the sources of plant suppliers and the identification of such plants. Nurseries are notoriously at fault when it comes to accurate plant identification.
5. Any alien plant removal projects should not involve herbicides. Only hand removal efforts can protect native species in border areas.

## Selected References

Cronquist, A. 1981. An integrated system of classification of flowering plants. Columbia University Press, New York, N. Y.

Dorn, R. D. 1988. Vascular plants of Wyoming. Mountain West Publishing, Cheyenne, Wyo.

Hartman, R., and R. Lichvar. 1980. Additions to the vascular flora of Teton County, Wyoming. Gr. Basin Nat. 40:408-413.

Kellogg, E. A. 1985. A biosystematic study of the *Poa secunda* complex. J. Arnold Arb. 66:201-242.

Marriott, H. 1991. Rare plants of Grand Teton National Park, second annual report. Prepared for GTNP by the Wyo. Nat. Diversity Database, The Nature Conservancy.

Reed, J. F. 1952. Vegetation of Jackson Hole Wildlife Park, Wyoming. Am. Midl. Nat. 48:700-729.

Shaw, R. J. 1958. Vascular plants of Grand Teton National Park. Am. Midl. Nat. 59:146-166.

————, 1968. Vascular plants of Grand Teton National Park, Wyoming revised. Sida 4:1-56.

________, 1976. Field guide to the vascular plants of Grand Teton National Park and Teton County, Wyoming. Utah State Univ. Press, Logan, Ut.

Spence, J. R., and R. J. Shaw. 1981. A checklist of the alpine vascular flora of the Teton Range, Wyoming, with notes on biology and habitat preferences. Gr. Basin Nat. 41:232-242.

Schmutz, E. M., and L. B. Hamilton. 1979. Plants that poison. Northland Press, Flagstaff, Az.

Whitson, T. D. et al. 1991. Weeds of the West. The Western Society of Weed Science in cooperation with the Western U. S. Land Grant Universities Cooperative Extension Services and the Univ. of Wyo.

Williams, M. C., and R. C. Barneby. 1977. The occurrence of nitro-toxins in North American *Astragalus* (Fabaceae). Brittonia. 29: 310–326.

# Index